The Complete Adventures of MR. TOMPKINS

IGOR GAMOW

SCORPIO STEELE

BIG BANG PRODUCTIONS, LLC

BOULDER

DEDICATION

To Stephen Jay Gould,
who opened the door
of evolutionary biology
to so many of us
struggling scientists,
via his magical book,
The Panda's Thumb.

THE COMPLETE ADVENTURES OF MR. TOMPKINS
Published by Big Bang Productions, LLC
Unless otherwise noted, all contents copyright ©2017 Big Bang Productions, LLC
All rights reserved.

No part of this publication may be reproduced, stored in a retrieval system, or transmitted, in any form or by any means, electronic, mechanical, photocopying, recording, or otherwise, without the prior written permission of the publisher, except in the case of brief quotations embodied in critical articles or reviews.

Introduction copyright ©2017 Paul D. Beale

Big Bang Productions, LLC
186 Cañon Park, Boulder, CO 80302

www.gamow.com

Printed in the United States of America.

ISBN: 978-1543054880

THE ADVENTURES OF MR. TOMPKINS
U.S. Trademark Registration No. 3941477
owned by Big Bang Productions LLC, H&H Ref. No. 74545.0002

CONTENTS

FOREWORD
BY PAUL BEALE

THE LEGEND OF MR. TOMPKINS
BY IGOR GAMOW
7

EINSTEIN'S NEW GRAVITY
9

UP AND ATOM
31

A ROUND WITH CURIE
57

ON THE ORIGIN OF SPECIES
91

OF PEAS AND GENES
123

DNA DISCOVERY
147

ARISTOTLE: THE BEST PURPOSE
175

LEONARDO DA VINCI
201

ON BEES AND SEAS
229

CREATED, MY CREATOR
255

AFTERWORD
BY IGOR GAMOW
295

FOREWORD

BY PAUL D. BEALE

George Gamow was a giant of science. He made seminal contributions to nuclear physics, astrophysics, and molecular biology. He was also an excellent teacher, and a prolific writer.

George was born in Odessa in the Ukraine in 1904. He studied physics, mathematics and astronomy at Novorossiya University in Odessa, and earned his doctoral degree at the University of Leningrad, studying with Alexander Friedmann until Friedmann's death in 1925. He was introduced to the leaders in quantum theory in Göttingen in 1928. While there, at the age of twenty-four, Professor Gamow was the first to use quantum mechanics to quantitatively explain the properties of radioactive alpha decay. This was the first application of quantum theory to nuclear physics. Remarkably, this was only two years after Werner Heisenberg and Erwin Schrödinger had developed quantum mechanics to explain the atomic structure of hydrogen. In the following few years Gamow developed theories of the thermonuclear reactions that power the sun and other stars, and first proposed the nuclear drop model that became instrumental in the theory of nuclear fission.

George and his wife Lyubov "Rho" Vokhminzeva Gamow defected to the West in 1933 after George was invited to attend the seventh Solvay conference that brought the greatest minds in physics together to discuss radioactivity and nuclear physics. Other attendees included Albert Einstein, Marie Curie, Ernest Rutherford, Louis de Broglie, Niels Bohr, Erwin Schrödinger, Werner Heisenberg, Wolfgang Pauli, Paul

Dirac, Enrico Fermi, Lise Meitner and many other leading physicists, including several additional future Nobel Laureates.

From 1934-1956 George was a professor of physics at George Washington University. George recruited Edward Teller to GW, and worked closely with him on the development of the theory of radioactive beta decay and other problems in nuclear physics. The annual conferences George established at GW, and later at the University of Colorado, attracted many leading scientists from around the world. It was at one of these conferences in 1939 that Niels Bohr announced to the scientific community the observation of nuclear fission by Hahn and Meitner.

In 1948 George and graduate student Ralph Alpher published a seminal paper on what became known as the standard model of Big Bang cosmology or, more colloquially, the Big Bang Theory. In their paper they calculated abundances of the light chemical elements created in the Big Bang 13.8 billion years ago. This work occurred in an era when the idea that the universe had begun abruptly was still highly controversial, yet Gamow and Alpher produced quantitative predictions based on the fundamental physics that follows from Edwin Hubble's astronomical observation that the universe is expanding. The paper was published on April 1, 1948. Perhaps not coincidentally, George, always the whimsical jester, added his friend Hans Bethe's name to the paper as a pun on the Greek alphabet. The paper is still widely known as the Alpher-Bethe-Gamow theory of primordial nucleosynthesis. In further work, Alpher and Robert Herman extended the theory and calculated that the universe should be filled with thermal microwave radiation left over from the Big Bang. This cosmic microwave background was discovered by Arno Penzias and Robert Wilson in 1963.

George's other very famous graduate student at George Washington was Vera Rubin. In her 1951 dissertation she reported the first observation of non-Hubble flow in the motion of galaxies. A decade later she and Kent Ford discovered dark matter by measuring the rotation curves of spiral galaxies.

In 1954, George turned his mind to explaining the genetic code. He and James Watson created the interdisciplinary RNA Tie Club that included Francis Crick, Max Delbrück, and other scientific leaders that helped unravel the mysteries of DNA. Gamow was the first to show that a three-letter nucleotide code was sufficient to define all of the known amino acids.

Professor Gamow was a prolific writer. His many successful books explained even the most arcane concepts of science and mathematics to budding scientists and the general public. George Gamow's *Mr. Tompkins* series introduced millions of people to physics and biology. The four books in the series—*Mr. Tompkins in Wonderland* (1939), *Mr. Tompkins Explores the Atom* (1944), *Mr. Tompkins Learns the Facts of Life* (1953), and with Martynas Ycas, *Mr. Tompkins Inside Himself* (1967)—set the standard for how scientists can effectively communicate the mysteries and beauty of science. This skill is reflected in his many other books. I still have the copy of *One, Two, Three... Infinity* that my parents bought for me when I was in junior high school. It is dedicated to, and kindly autographed by, George's son Igor Gamow, "who wanted to be a cowboy." In recognition of the importance of his popularization of science, the United Nations Educational, Scientific, and Cultural Organization (UNESCO) awarded George Gamow the Kalinga Prize in 1956.

That same year George Gamow joined the faculty of the Department of Physics at the University of Colorado. He quickly became a campus leader, and his teaching and writing influenced a new generation. He published over a dozen additional textbooks and popular books on science. *My World Line: An Informal Autobiography* (1970) tells George' incredible life story, and the final chapter in this collection, *The Adventures of Mr. Tompkins and George Gamow,* is based upon this book.

George Gamow died in 1968, but his legacy continues. His campaign to create a modern physics building at CU came to fruition, and the office tower of the Duane Physical Laboratories bears his name. In 1971 the Department of Physics established the George Gamow Memorial Lecture series. Four years later George's second wife, Barbara "Perky" Gamow, left a bequest to the University to permanently fund the public lecture series. The Gamow Memorial Lecture endowment stipulates that the Gamow lecturer "shall be a person widely known for scholarly achievement in one or more of the fields of physics, astrophysics, mathematics, biology, chemistry, geology or engineering, …[and] shall be known for his or her excellence in interpreting science for the general public." The university has hosted fifty George Gamow Memorial Lectures, and the lecturers have been many of the leading scientists of the twentieth and twenty-first centuries, including twenty-five Nobel Laureates. Recent lecturers include Jane Goodall, Adam Reiss, Frank Wilcek, and Brian Greene. In 2017 the fifty-first Gamow lecturer will be Kip Thorne, who gratefully agreed. "George Gamow," Thorne says, "played a huge role in getting me hooked on physics when I was a teenager." It has been my honor to serve as the chair of the George Gamow Memorial Lecture Committee since 2008. I have become close friends with Igor Gamow and his wife, Elfriede Reger Gamow, who have graciously supported the lecture series, and have created a physics scholarship in their names.

George Gamow has had a profound impact on science, and on the public dissemination of scientific knowledge. Igor's continuation of the *Mr. Tompkins* series pays homage to his father and adds to his legacy. I am confident that *Mr. Tompkins* will continue to inspire future generations.

Paul D. Beale
Professor of Physics
University of Colorado Boulder

Professor Paul D. Beale is a professor of physics at the University of Colorado Boulder. A theoretical physicist specializing in the statistical mechanics of condensed matter systems, his research interests include exact solutions of statistical mechanical models, the thermodynamics, phase transitions, and critical properties of materials, and pseudorandom number generators. With R.K. Pathria he coauthored the third edition of Statistical Mechanics, *a graduate physics textbook.*

Mr. Tompkins is my brother and our father is George Gamow. As you just learned in Paul's introduction, he is also revered as the creator of a series of popular science books starring the eponymous hero, Mr. Tompkins. While the final story in this book presents a mythologized account of Tompkins' creation, I present to you his true origin.

The first time the name "Tompkins" left Father's lips was in the summer of 1934, in Ann Arbor, Michigan. Father was a visiting professor there, and when he arrived he met his first two American scientists—Mr. Kemble*, a physics graduate student who took care of Father's professional and personal needs, and Mr. Tompkins, a mathematician whose first name has unfortunately been lost to the sands of time. Father, a Russian, thought it was the most interesting and unusual name he had ever heard and stored it away in his memory.

A few years later, in 1938, Father decided to write a story about a bank clerk who was completely ignorant of science. Bored with celebrity gossip in the entertainment page of the newspaper, the clerk decided to attend a popular-science lecture on Einstein's theory of relativity at the local university. As the tall, white-bearded Professor lectured, the clerk fell asleep and had wild dreams in alternate worlds in which the Professor taught him science. Father remembered the Tompkins of Ann Arbor and christened the bank clerk "Mr. Tompkins". As he could not remember the first name of the real-life Tompkins, he gave his character the initials "c. G. h", from the three fundamental constants in physics (c for the speed of light, G for gravity, and h for Planck's constant, which describes the sizes of quanta in quantum mechanics.

He submitted the first story to half a dozen magazines, including *Harper's Magazine* and *The Atlantic Monthly*, but it was

---

*Father came to call Kemble his "batman". The modern reader might immediately conjure a mental image of the comic book superhero, but this old military term derives from a combination of the French *bât*—"packsaddle"—and "man". In the British army each officer was assigned an orderly, a *batman*, who would take care of all his needs, from doing laundry to shining shoes. (I do not know about the laundry, but I'm pretty sure Mr. Kemble did not shine Father's shoes!)

rejected. Rather uncharacteristically, Father decided to abandon the Tompkins project.

The following year he was in Warsaw, Poland, for an astronomy conference. There he met British physicist Sir Charles Galton Darwin, grandson of *the* Charles Darwin, the famous naturalist credited with the concept of natural selection. Over a dinner held during the conference, Father told his Tompkins story to Sir Charles, who very much liked it and suggested he send a copy to C.P. Snow. Snow had written some detective stories and, more importantly, was the editor of *Discovery* magazine, published by Cambridge University Press. Snow loved the first Tompkins story and asked for more, which he soon serialized. Cambridge then asked Father to expand his stories to book length, which the university published in 1940 as *Mr. Tompkins in Wonderland*. Over the years more books followed, such as *Mr. Tompkins Explores the Atom* and *Mr. Tompkins Learns the Facts of Life.*

During the forty-odd years that I was a professor at the University of Colorado, dozens of visiting scientists would pass my door, see my name, and ask me the following two questions:

"Are you the son of George?"
"Yes," I answered, "According to my mother."
"Did you become a cowboy, as your father said you "would rather be" in *One, Two, Three... Infinity*?"
My answer: "Yes, of course."

In almost every case these visitors informed me that they had gone into science because they were inspired by my father's books, so the goal of the book you hold in your hands is to continue to inspire people, children and adults alike, to go into science.

Although the original versions of the Mr. Tompkins books are now out of print, there are many updated Tompkins books available, some featuring additional coauthors. Since I have always been a fan of comic books I have, with the help of artistic collaborator Scorpio Steele, adapted Tompkins into sequential art, where Tompkins encounters some of the most famous scientists in history.

Two of the scientists in this book were chosen because of the following story. Both of my parents escaped Stalin's Soviet Union and ended up in Cambridge, England. When I was a boy, Father enjoyed telling all who would listen about how he and my mother, Rho, had borrowed money from the three most famous physicists in the world: Marie Curie, Ernest Rutherford, and Niels Bohr. After receiving his first paycheck in America, Father wired the loaned money back to Europe. But he always regretted not keeping the receipts inscribed with those three famous autographs.

Father felt that most science books usually ignore the scientists who originated a concept or discovered something, except for the obligatory dates of birth and death. On the other hand, biographies of the scientists, for the most part, ignore the science. My passion and goal for this series is to, like Lewis Carroll's *Alice's Adventures in Wonderland* and *Through the Looking Glass*, combine both the visually delightful and the intellectually stimulating into a format people of all ages can enjoy for years to come.

One final note: the first four stories in this book were originally created to be the second half of a video episode, and refer to information presented in the videos. While these stories are enjoyable and informative in and of themselves, a deeper understanding of the scientific concepts may be had by first watching the videos. The URL addresses for each video are listed on the first page of the story.

R. Igor Gamow
Retired Professor
Chemical and Biological Engineering
University of Colorado Boulder

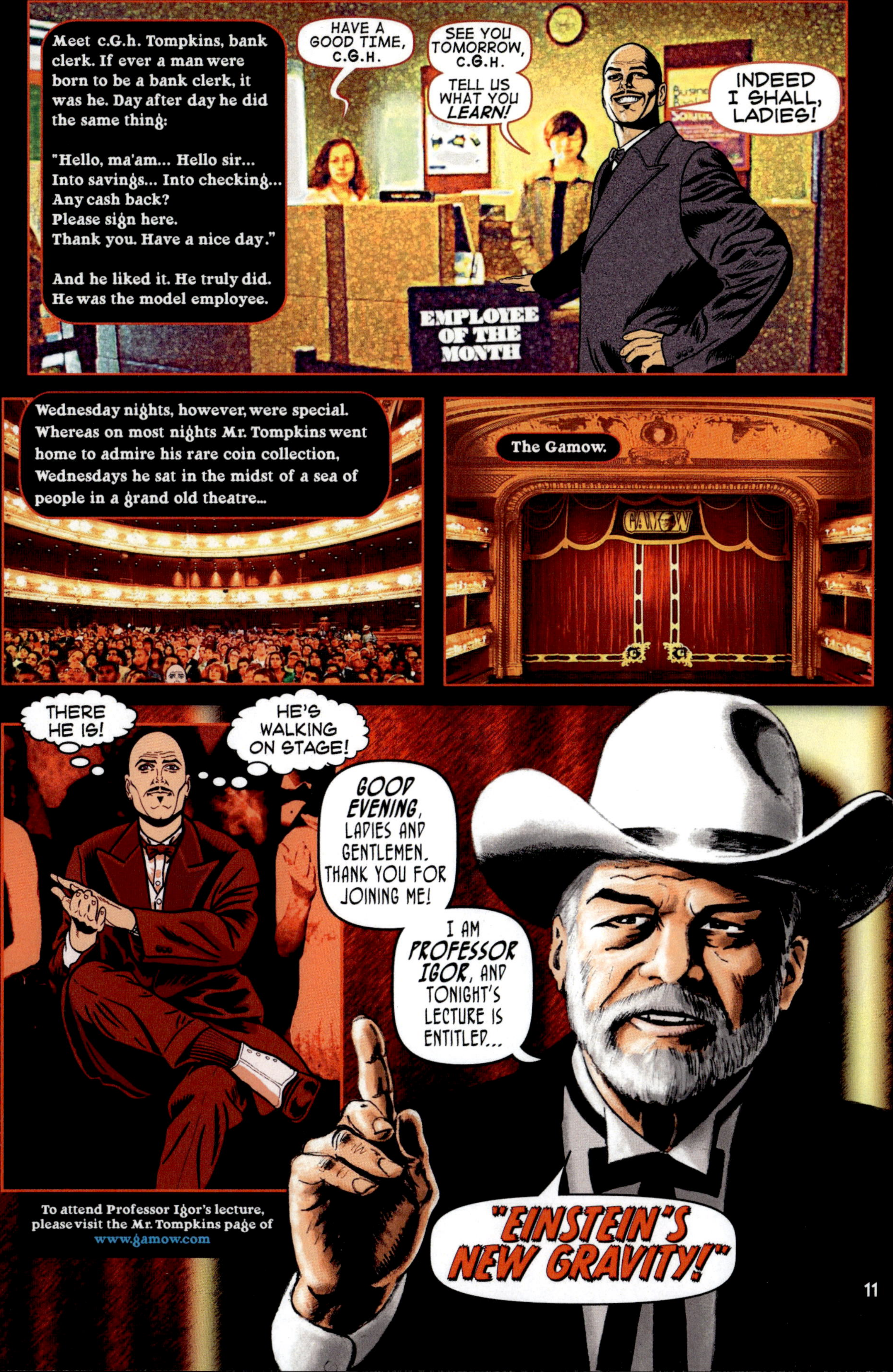

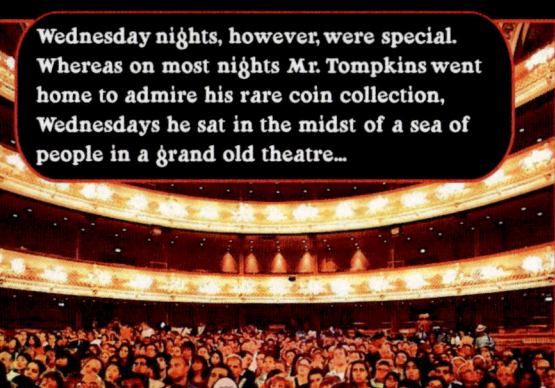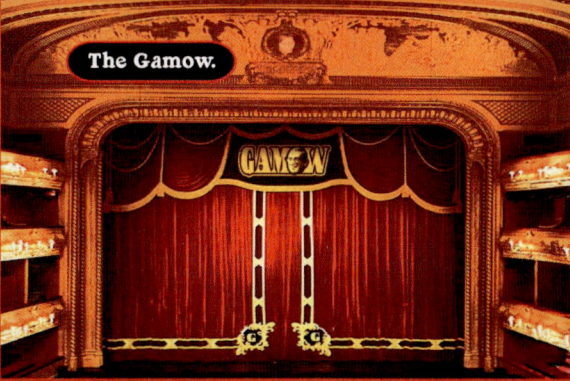

"The most beautiful thing we can experience is the mysterious. It is the source of all true art and science. He to whom this emotion is a stranger, who can no longer pause to wonder and stand rapt in awe, is as good as dead: his eyes are closed."

Albert Einstein
Living Philosophies

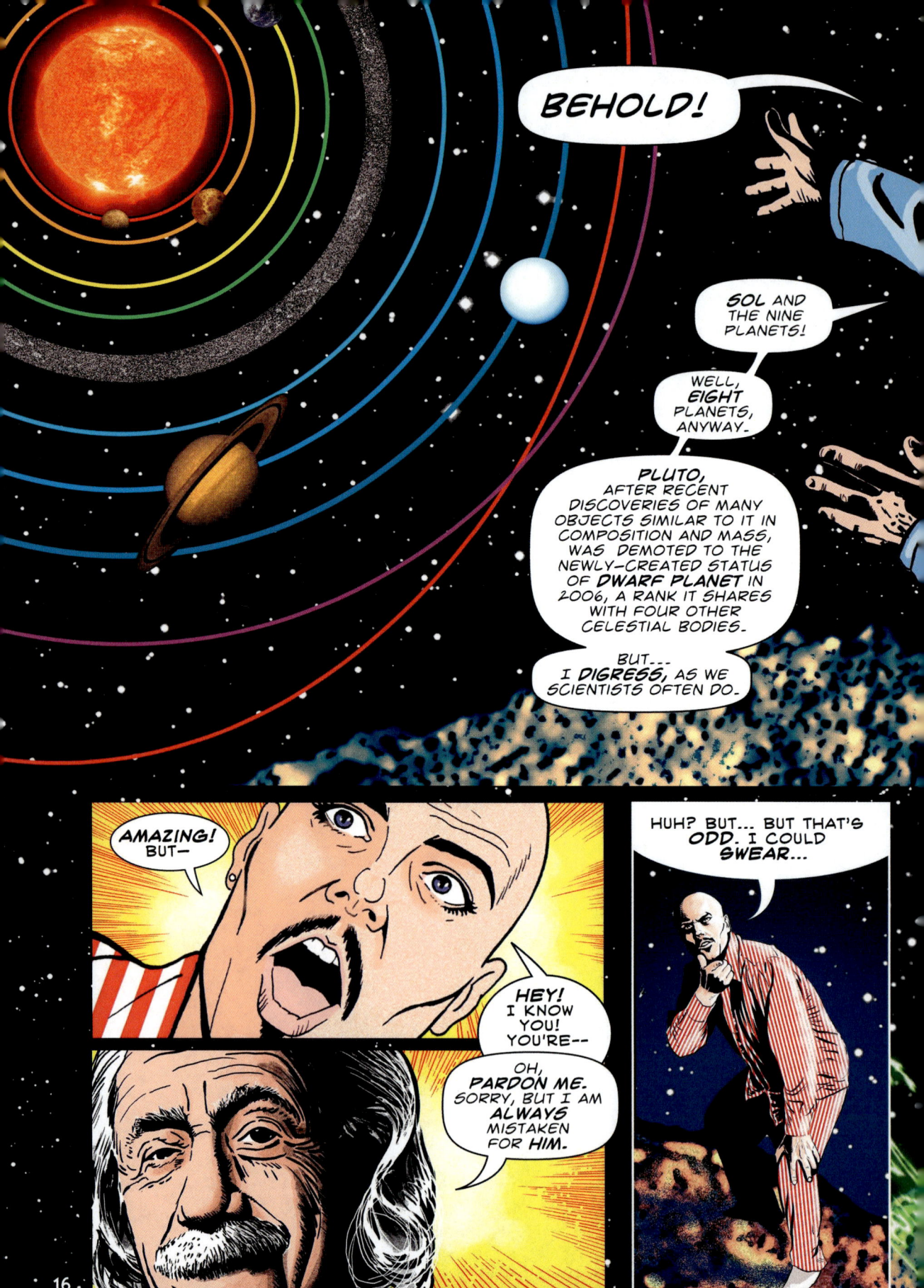

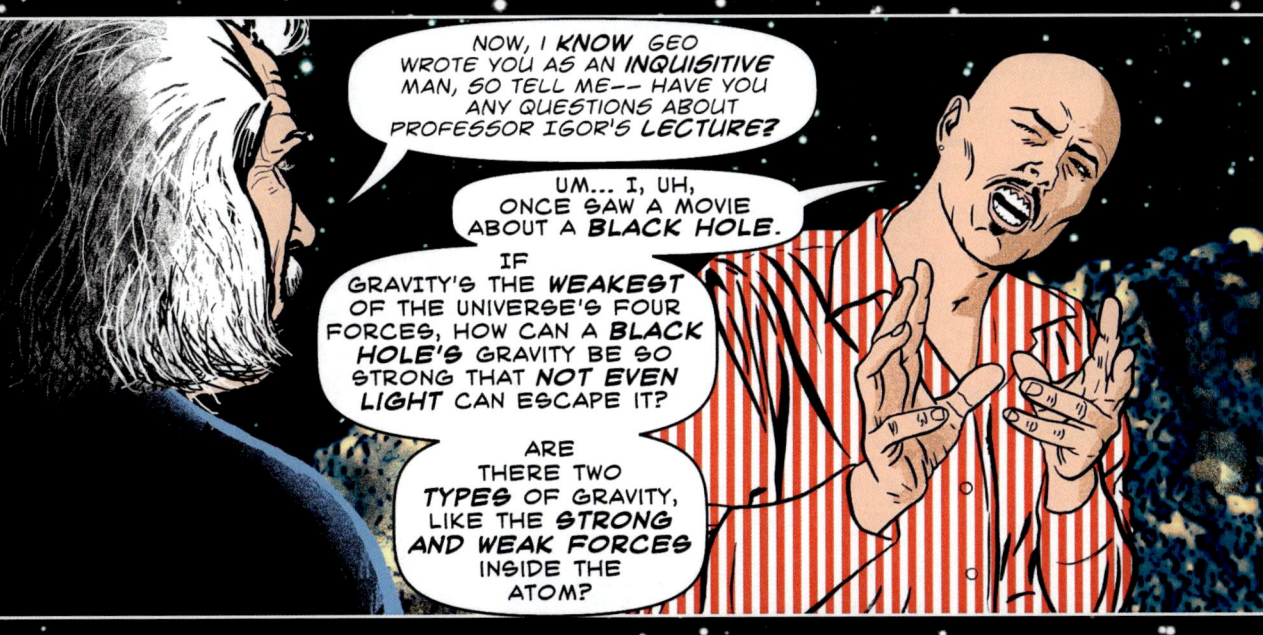

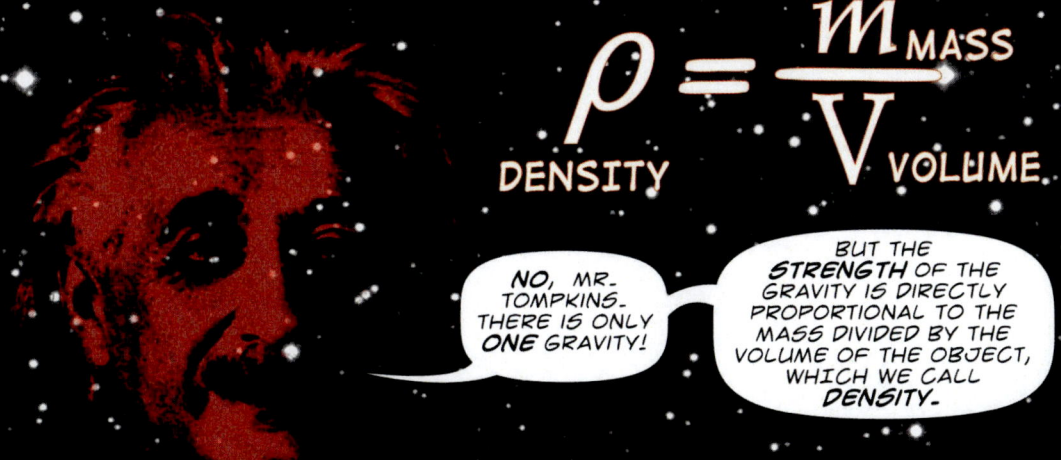

"WATCH CLOSELY, MY BOY, AS I PLUCK EARTH FROM ITS ORBIT, LIKE SO...

...SQUISH IT DOWN TO THE SIZE AND VOLUME OF A CHERRY, THUS INCREDIBLY INCREASING ITS DENSITY...

...THEN SHRINK YOU TO THE SIZE OF AN ANT!"

ARRGGHH!

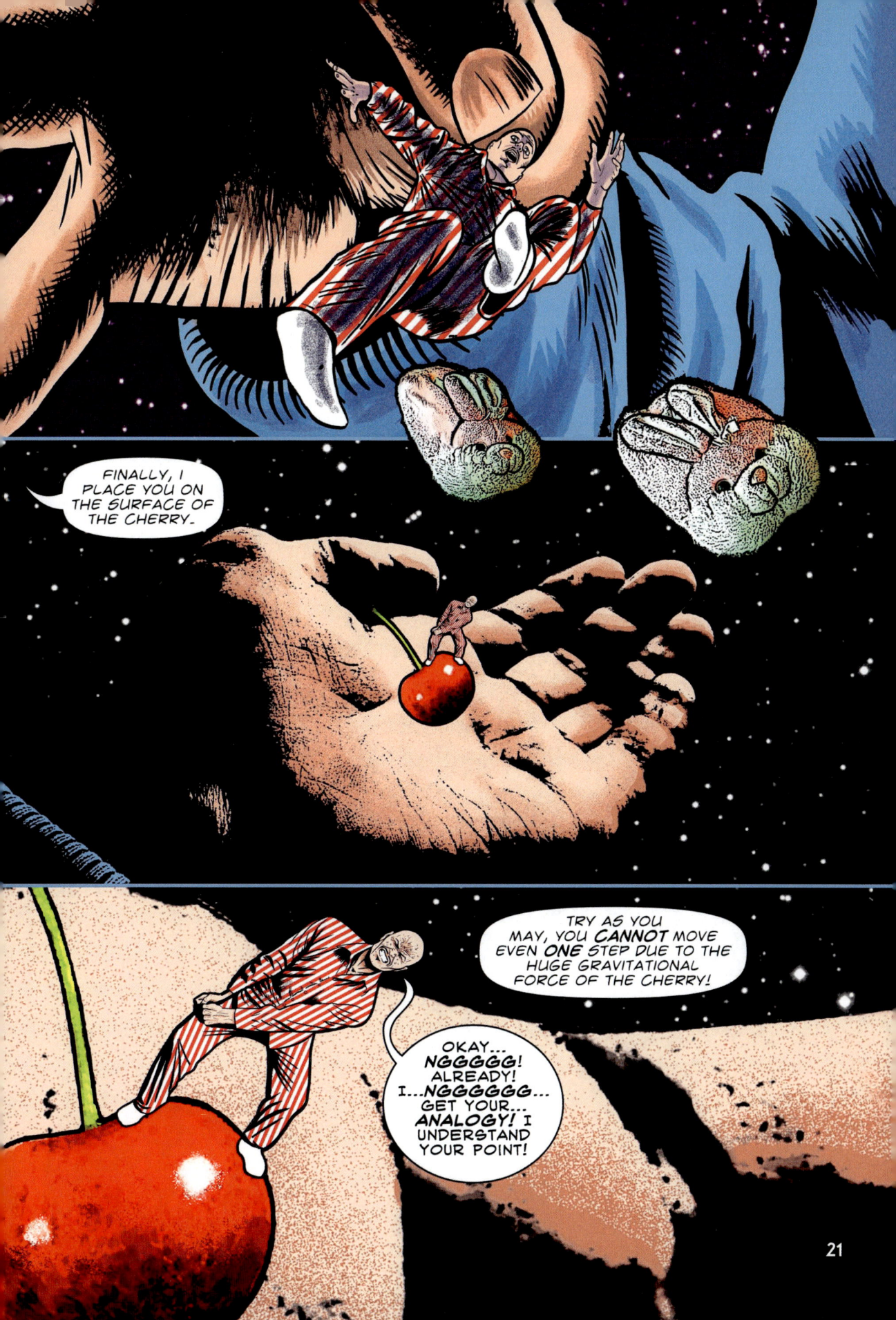

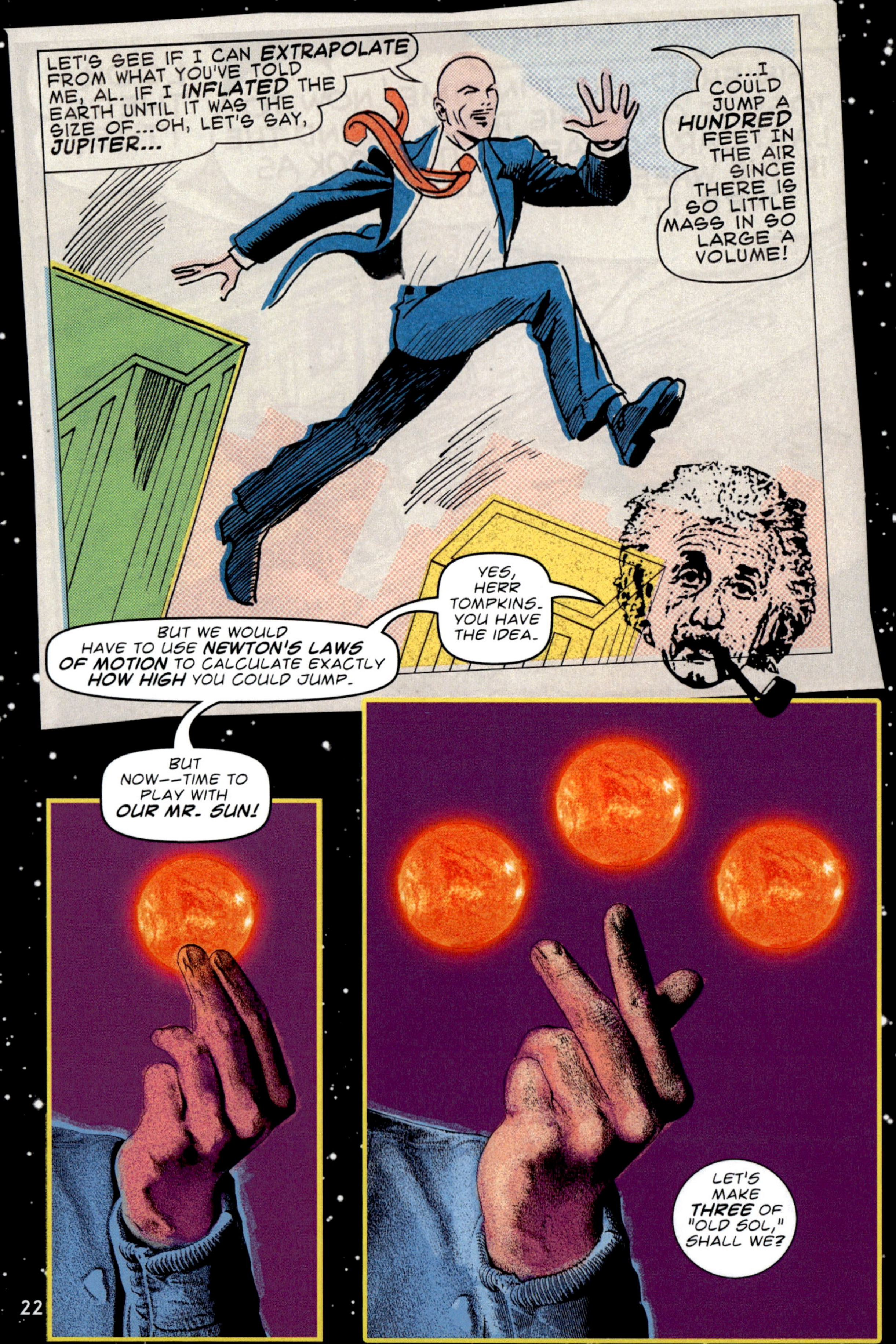

"I, HOWEVER, HAVE CLEARLY SHOWN THAT NOTHING CAN GO FASTER THAN LIGHT-SPEED!"

"FOR EXAMPLE, NEWTON BELIEVED..."

"...THAT IF THE EARTH SUDDENLY MOVED 1000 MILES CLOSER TO OUR MOON, LUNA WOULD ALTER ITS ORBIT INSTANTANEOUSLY."

"I, EINSTEIN! PREDICT A SIGNIFICANT DELAY BEFORE THE MOON ALTERED ITS ORBIT!"

"ON A LARGER SCALE, IF THE SUN WERE SUDDENLY TO DISAPPEAR, NEWTON WOULD PREDICT THAT THE PLANETS WOULD IMMEDIATELY FLY OFF INTO SPACE AT THE EXACT MOMENT THE SUN VANISHED."

"COULD THE SUN SUDDENLY DISAPPEAR? JUST...ONE MINUTE IT'S HERE, THE NEXT IT'S GONE?"

"NO, IT COULDN'T, BUT FOR THE SAKE OF OUR GEDANKEN EXPERIMENT, LET US SAY IT COULD!"

"SINCE GRAVITY WAVES TRAVEL AT THE SAME SPEED AS LIGHT-- 186,000 MILES PER SECOND...

"IT WOULD THUS TAKE **EIGHT MINUTES** BEFORE WE ON EARTH WOULD SEE THE SUN **'BLINK OUT!'**

"**EIGHT MINUTES** UNTIL DARKNESS ENVELOPED THE EARTH!"

"**EIGHT MINUTES** UNTIL EARTH CAREENED OFF INTO SPACE!"

IF EARTH LEFT ORBIT WOULD PEOPLE FLOAT OFF AS IF THEY WERE IN THE "VOMIT COMET"?

NO. **EARTH'S GRAVITY**, FOR ALL PRACTICAL PURPOSES, IS **INDEPENDENT** OF THE SUN'S GRAVITATIONAL PULL.

THEY **WOULD**, HOWEVER, NOTICE THE **STARS** IN NEW, **UNFAMILIAR** POSITIONS, AND THE SOLAR SYSTEM'S PLANETS AND MOONS WOULD BE **INVISIBLE** TO US WITHOUT SUNLIGHT REFLECTING OFF OF THEM.

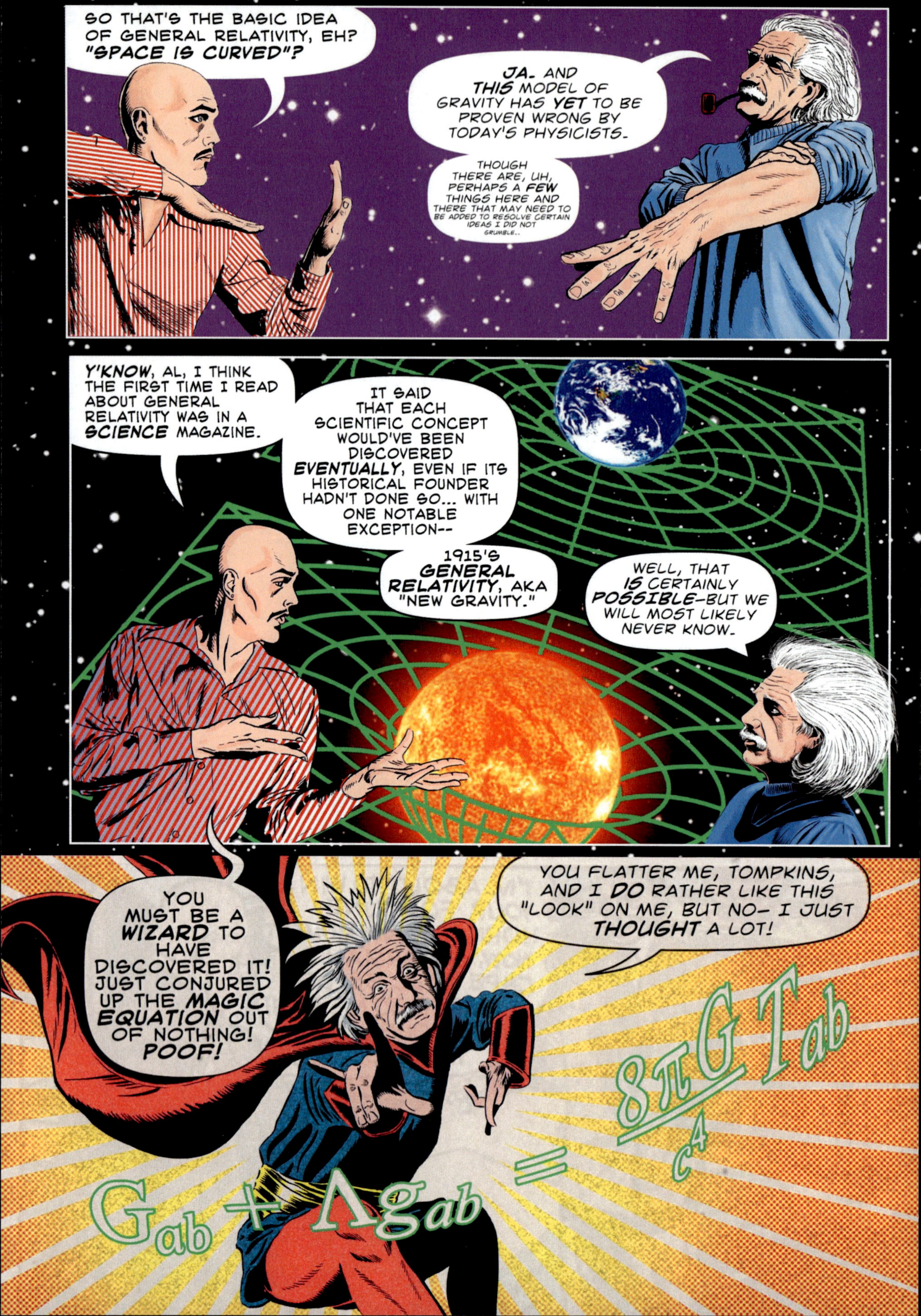

"Gentlemen, now you will see that now you see nothing. And why you see nothing you will see presently."

Sir Ernest Rutherford

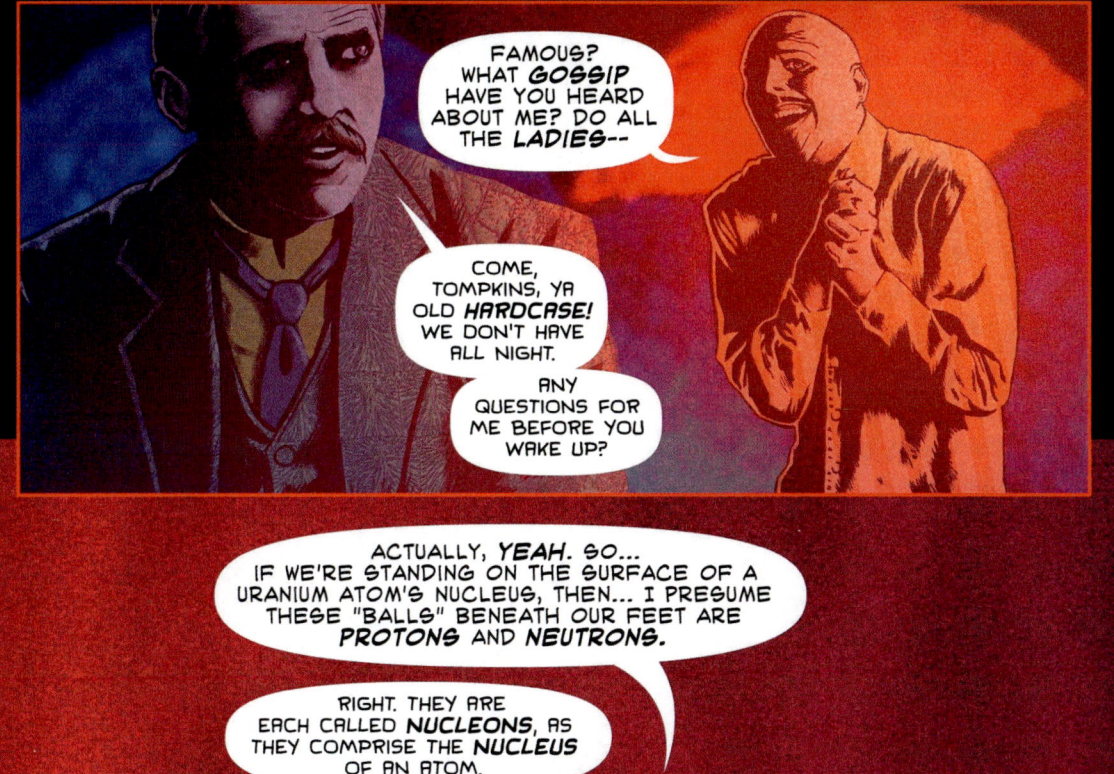

BUT BEFORE THIS, I FIRST HAD TO DISCOVER THE **NUCLEUS ITSELF!** HERE'S HOW I DID IT!

GEIGER

MARSDEN

PSST! AN ALPHA PARTICLE IS A POSITIVELY CHARGED NUCLEAR PARTICLE MADE OF TWO PROTONS AND TWO NEUTRONS.

AND... IT'S THE SAME THING AS AN *HELIUM NUCLEUS!*

IN 1909, UNDER MY DIRECTION, **HANS GEIGER** AND **ERNEST MARSDEN** PERFORMED WHAT BECAME A VERY *FAMOUS EXPERIMENT!*

FROM SOME RADIOACTIVE *RADIUM* WITHIN A LEAD BOX, WE AIMED A STREAM OF *ALPHA PARTICLES* THROUGH A HOLE IN THE BOX...

SOME ALPHA PARTICLES CAREENED OFF WILDLY, AND A FEW EVEN *BOUNCED BACK AT THE BOX!*

IT WAS ALMOST AS INCREDIBLE AS IF YOU FIRED A *15-INCH SHELL* AT A PIECE OF *TISSUE PAPER* AND IT CAME BACK AND HIT YOU!

THIS IS WHEN I FIGURED THAT ONLY A HEAVY, POSITIVELY CHARGED PARTICLE, SUCH AS THE PROPOSED *NUCLEUS*, COULD ACCOUNT FOR THE BACKWARDS RICOCHET.

e-

+

I FIGURED THAT THE NEGATIVELY CHARGED ELECTRONS, *BALANCED* BY THE POSITIVE CHARGE OF THE NUCLEUS, MIGHT *ORBIT* THE NUCLEUS LIKE THE *PLANETS ORBIT THE SUN*, THE ELECTROSTATIC FORCE OF ATTRACTION AKIN TO *GRAVITY.*

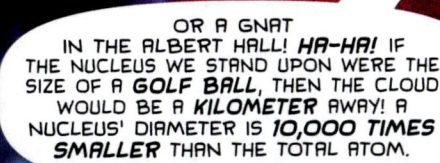

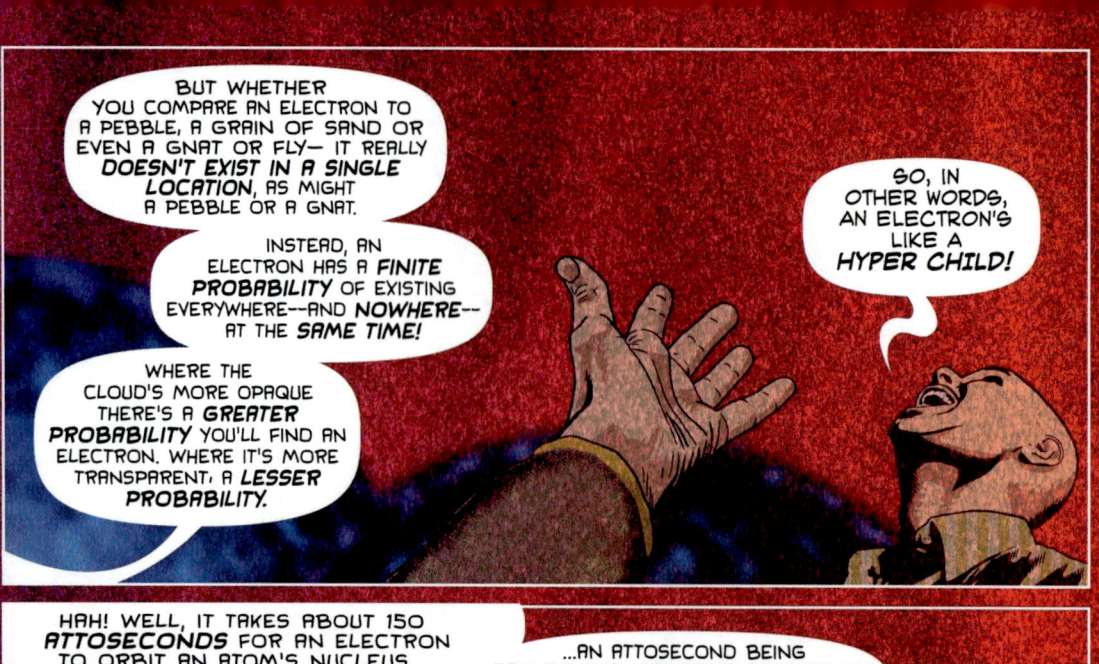

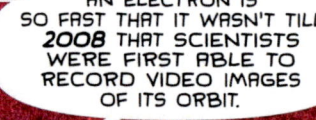

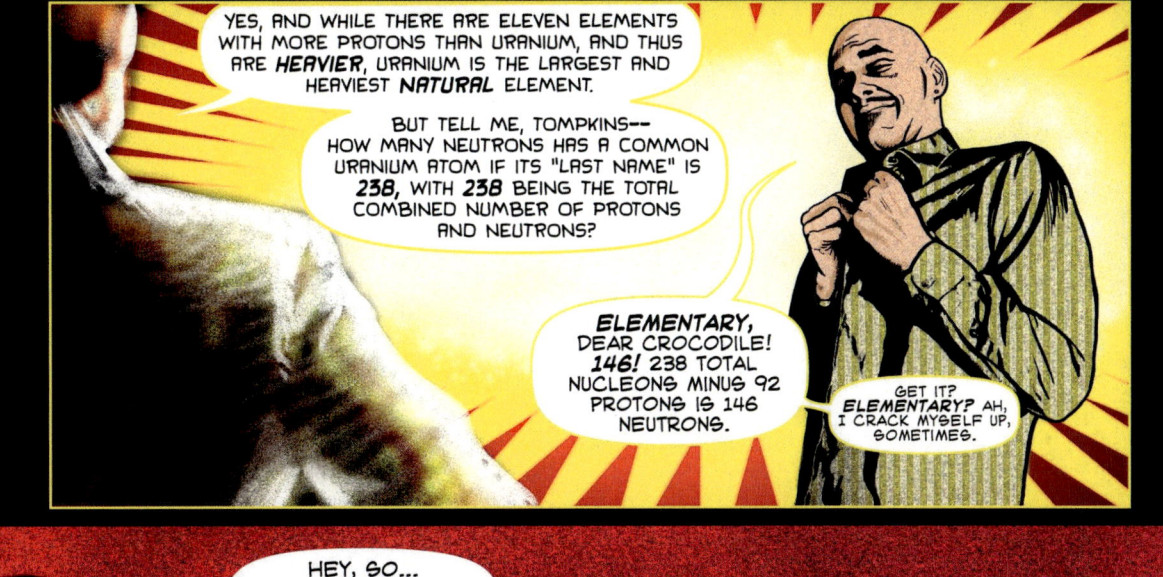

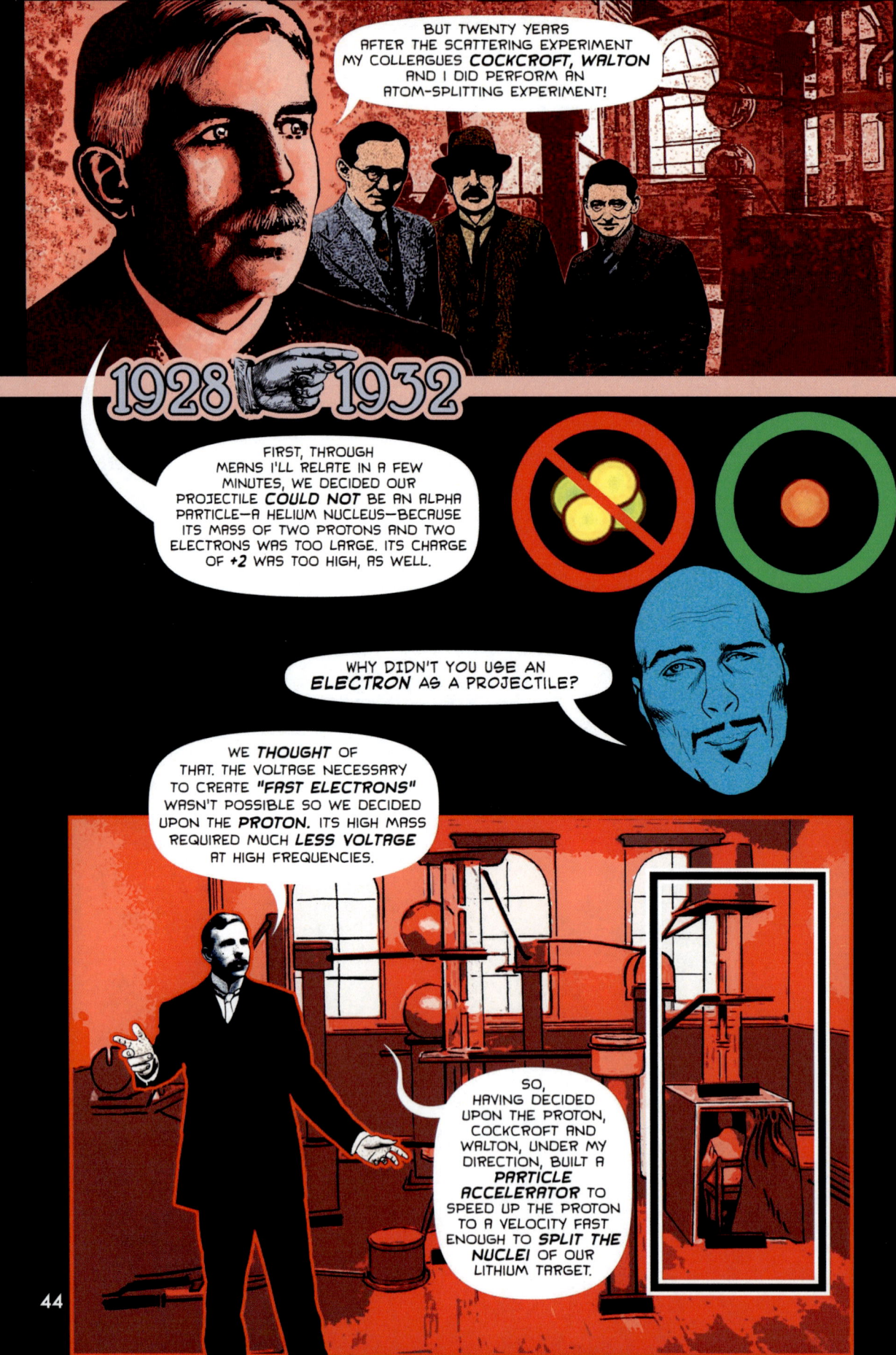

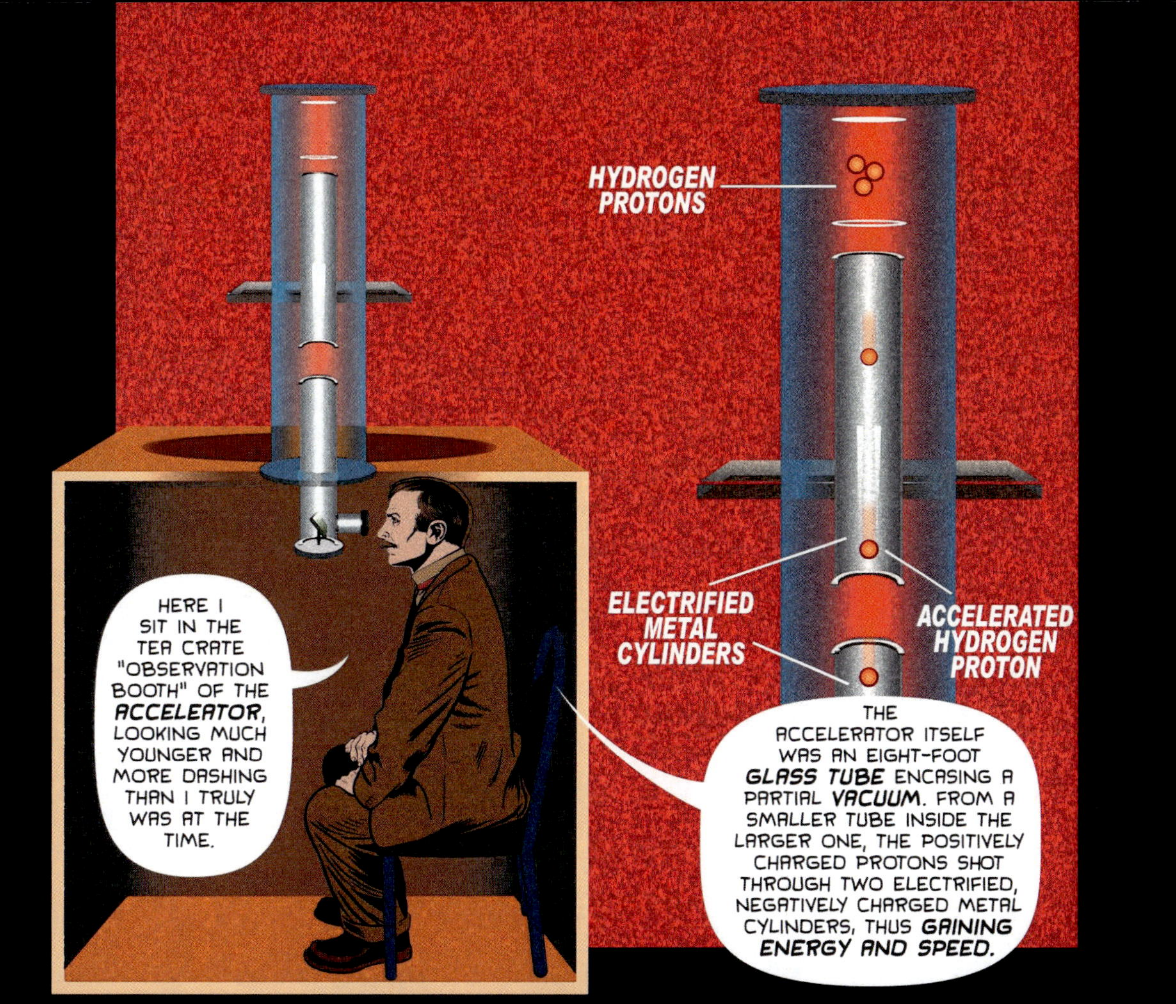

HYDROGEN PROTONS

ELECTRIFIED METAL CYLINDERS

ACCELERATED HYDROGEN PROTON

"HERE I SIT IN THE TEA CRATE "OBSERVATION BOOTH" OF THE *ACCELERATOR*, LOOKING MUCH YOUNGER AND MORE DASHING THAN I TRULY WAS AT THE TIME."

"THE ACCELERATOR ITSELF WAS AN EIGHT-FOOT *GLASS TUBE* ENCASING A PARTIAL *VACUUM*. FROM A SMALLER TUBE INSIDE THE LARGER ONE, THE POSITIVELY CHARGED PROTONS SHOT THROUGH TWO ELECTRIFIED, NEGATIVELY CHARGED METAL CYLINDERS, THUS *GAINING ENERGY AND SPEED*."

"WHEN THE PROTONS LEFT THE CYLINDERS THEY STRUCK THE LITHIUM TARGET AT THE BOTTOM OF THE TUBE. WE USED LITHIUM FOR THE TARGET BECAUSE IT'S THE *LIGHTEST* KNOWN METAL ELEMENT ON THE PERIODICAL TABLE, WITH TWO PROTONS AND ONE NEUTRON."

LITHIUM TARGET

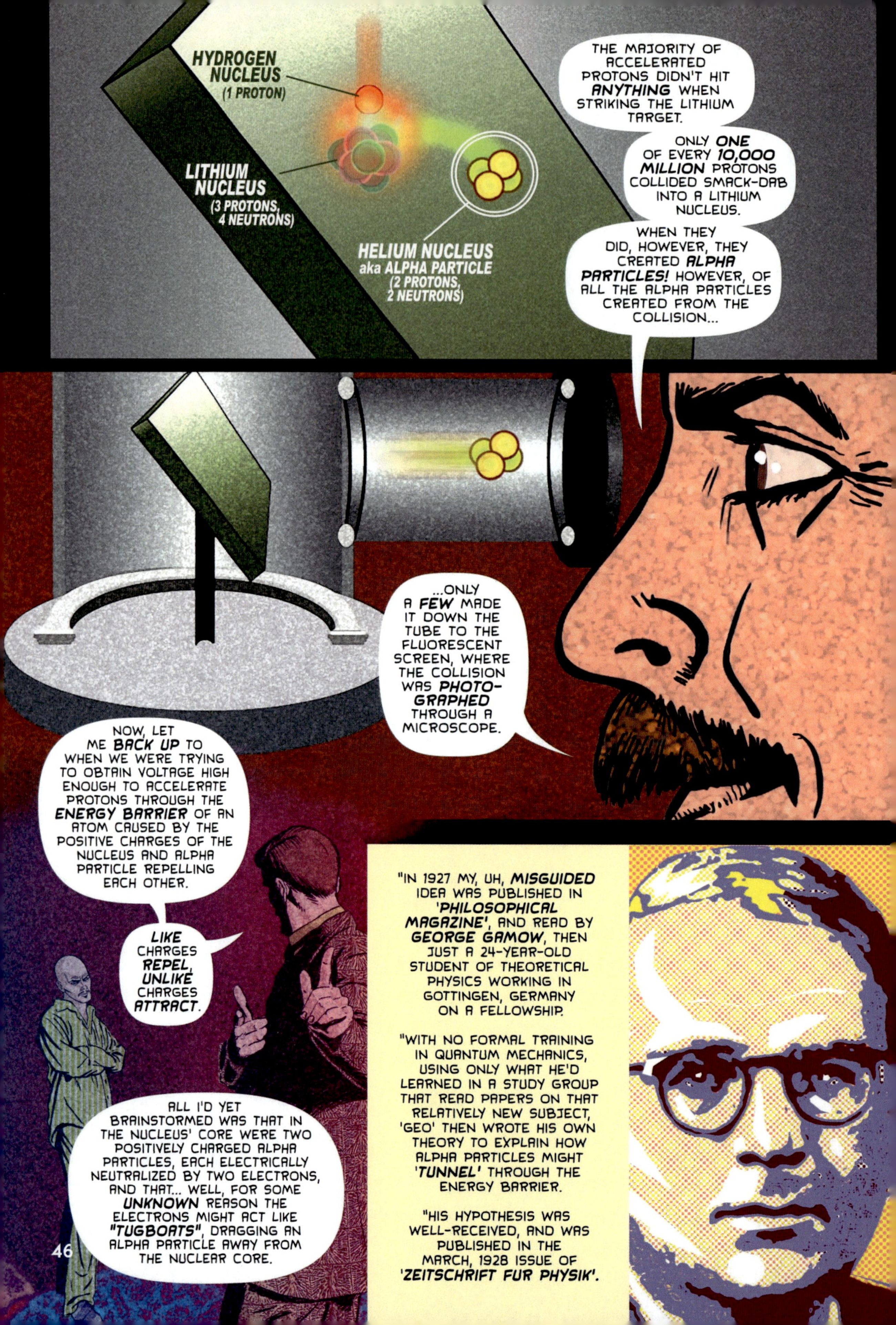

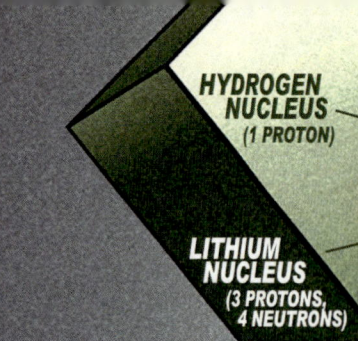

"GEO'S GERMAN FELLOWSHIP WAS AT AN END, THOUGH, SO ON HIS WAY HOME TO LENINGRAD, HE STOPPED IN COPENHAGEN, TO MEET NOBEL-PRIZE-WINNING PHYSICIST **NIELS BOHR** AT HIS INSTITUTE OF THEORETICAL PHYSICS..."

"BOHR WAS SO IMPRESSED WITH GEO'S CALCULATIONS THAT HE ARRANGED FOR HIM A YEAR-LONG **CARLSBERG FELLOWSHIP** AT THE **ROYAL DANISH ACADEMY OF SCIENCES**. WHEN THAT ENDED, GEO CAME TO ME IN CAMBRIDGE WITH A LATTER OF INTRODUCTION FROM BOHR."

"COCKCROFT ADAPTED GEO'S CALCULATIONS FOR PROTON USE THEN EXPLAINED THAT OUR ACCELERATOR WOULD ONLY NEED **300,000** VOLTS, RATHER THAN THE **8 MILLION** I'D PROPOSED. THOUGH I WAS A BIT WARY OF THEORETICAL PHYSICISTS, PREFERRING SOLID **LAB WORK** TO PURE MATH, I HAD READ GEO'S ARTICLE AND REALIZED ITS SIGNIFICANCE.

COCKCROFT   GAMOW

"BEFORE WE BUILT THE ACCELERATOR GEO VISITED US IN **JANUARY, 1929**, REINFORCING TO ME THE VALUE OF COCKCROFT'S PLAN. COULDN'T HAVE DONE IT WITHOUT GEO!"

I SUPPOSE THE FACT THAT MY CREATOR WAS SO INTELLIGENT EXPLAINS A LOT ABOUT **ME**, EH?. HEH, HEH, HEH.

BUT TELL ME THIS, CROCODILE—

IF WE'RE ON TOP OF THE NUCLEUS OF A URANIUM ATOM MIGHT WE GET BLOWN TO SMITHEREENS IF IT DECIDES TO DISINTEGRATE INTO ANOTHER KIND OF URANIUM ATOM?

WELL, **MISTER GENIUS**, IT COMES DOWN TO MY CONCEPT OF THE **HALF-LIFE**— THE TIME IT TAKES ONE HALF OF A RADIOACTIVE MATERIAL'S ATOMS TO DISINTEGRATE.

92 U 238

(almost)

HALF-LIFE

4.47 Billion Years

THE HALF-LIFE OF **92 U 238** IS ABOUT **4.5 BILLION YEARS**— NEARLY THE SAME AGE AS THE EARTH.

THUS, THE CHANCE THAT OUR URANIUM ATOM WILL EXPLODE IN THE NEXT FEW MINUTES IS **EXTREMELY** REMOTE.

4.55 Billion Years

I'M A BIT **EMBARRASSED** TO ADMIT I MADE QUITE A DUMB STATEMENT IN 1933, WHEN I SAID...

"THE ENERGY PRODUCED BY THE BREAKING DOWN OF THE ATOM IS A VERY **POOR** KIND OF THING. ANYONE WHO EXPECTS A SOURCE OF **POWER** FROM THE TRANSFORMATION OF THESE ATOMS IS TALKING **MOONSHINE.**"

OH, WELL. YOU LIVE AND YOU LEARN.

PTTTT!

"SLOW" NEUTRON

ANYHOO, LET'S SAY WE WANT TO SPLIT AN **URANIUM-235** ATOM, THE SECONDMOST COMMON VARIETY OF URANIUM*

URANIUM-235 IS A BIT **LESS STABLE** THAN URANIUM-238 (92 U 238), WHICH MAKES IT **FISSILE**, OR SPLITTABLE.

TO SPLIT IT WE USE A **"SLOW" NEUTRON**, WHICH HAS LESS **KINETIC ENERGY** THAN A "FAST NEUTRON", AND IS THUS LESS LIKELY TO BOUNCE OFF OF THE NUCLEUS.

DUE TO ITS PROTONS, THE ATOMIC NUCLEUS HAS AN OVERALL **POSITIVE** CHARGE. BUT SINCE THE NEUTRON IS ELECTRICALLY **NEUTRAL**, WHEN IT STRIKES THE NUCLEUS IT IS NOT REPELLED, BUT **ABSORBED** INTO IT.

URANIUM-235 NUCLEUS

*IN THE NEXT CHAPTER, **MADAME CURIE** WILL EXPLAIN **ISOTOPES**, THE DIFFERENT VARIETIES OF AN ATOM

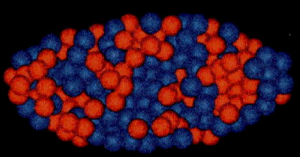

THIS CHANGES URANIUM-235 INTO THE VERY UNSTABLE **URANIUM-236**...

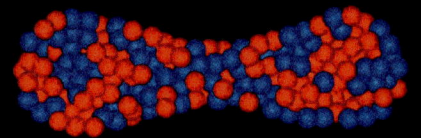

WHICH SOON DECAYS...

BARIUM NUCLEUS

KRYPTON NUCLEUS

"FAST" NEUTRONS

...AND **SPLITS!** THIS RELEASES ENERGY AND CREATES TWO NEW ELEMENTS, **BARIUM** AND **KRYPTON**, PLUS THREE "FAST" NEUTRONS.

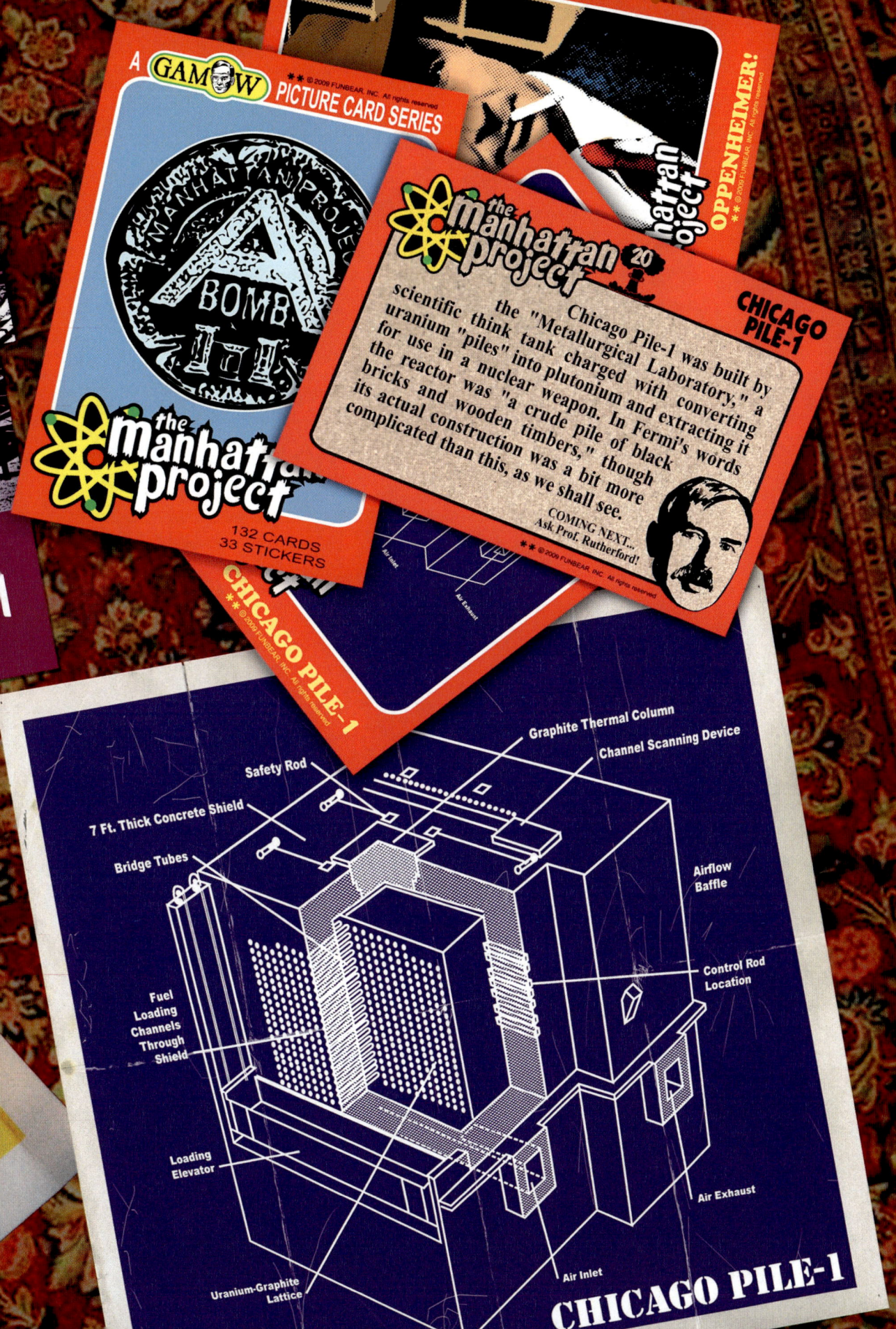

THE ADVENTURES OF MR. TOMPKINS

ASK "Prof." Rutherford

Dear Prof. Rutherford:
 How was "Chicago Pile-One" constructed?

C.G.H. Tompkins
Boulder, CO

*Good question, Mr. Tompkins. The wood provided scaffolding for a lattice of over 12,000 pounds of pure **uranium metal** and another 84,000 pounds of powdered **uranium oxide**. Stacks of black **graphite bricks**, two-thirds of which were imbedded with uranium **"pseudospheres"** containing both U235 and U238, were used for slowing the neutrons down to a ___ more conducive to fission.*

Original architect's sketch of the pile under construction.

Level 3

Level 19

Level 29

Chicago Pile-One in various stages of construction.

As neutrons shoot through the brick they collide against the carbon atoms in the graphite, which act rather like sandbags do for bullets on a firing range. These slowed-down neutrons have little chance of penetrating U238 but have a much greater probability of penetrating U235.

The effect, as we have seen, is that with a huge release of energy, U235 splits into the "daughter" elements barium and krypton with additional fast, footloose and fancy-free neutrons which careen into other U235 atoms, splitting them and continuing the chain reaction.

To prevent an uncontrolled fission chain reaction occurring within the pile, Fermi and his team inserted cadmium rods into it to absorb the free neutrons.

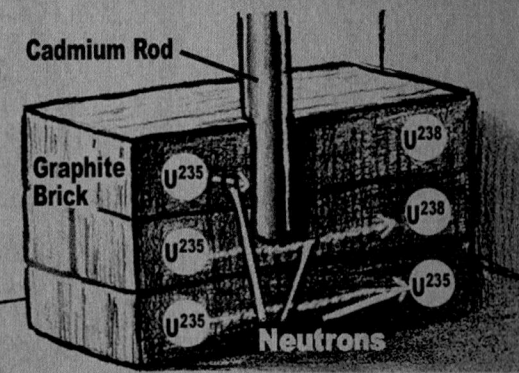

MR. TOMPKINS EXPLORES THE ATOM

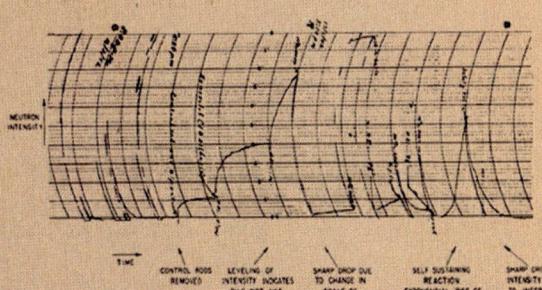

Enrico Fermi (front, far left) and the CP-1 Team

BLOOD PRESSURE GAUGE

for HOME USE.......... $5.98

Unsuspected rises in blood pressure is one of the most important single causes of disability and death in the United States today. Heart trouble, brain strokes, kidney damage, and other critical ailments can be avoided if you are able to keep constant check on the condition of your health and blood pressure. This simple-to-operate instrument is dependable and ideally suited for home use. You do not have to have a partner because you can take your own blood pressure. Complete operating instructions are included as well as information on what doctors advise, what you should do, and when to do it.

Those already having high blood pressure can find relief from undue worry over their condition, when they are able to check their own pressure at home. Then they'll know just when to see the doctor.

This complete unit includes a pressure gauge, 25 grams of mercury, heavy vinyl bandage, double valved bulb, stethoscope, base, and manual. Additional information upon request. DO THIS NOW: Send just your name and address and pay $5.98 plus C.O.D. charges on arrival. Cash orders shipped postpaid.

JOY SPECIALTY COMPANY
2320-Z West Hubbard Street   Chicago 12, Illinois

12 SETS—PIN-UP GIRL-PHOTOS $1.00

Here is the Largest, most spectacular collection of Beautiful Pin-up Girl Photos you have ever seen. You will find mostly every type of girl pose, from a Bathing Beauty to Beautiful Art Models and Show Girls. Never before, a collection of this kind was ever sold at such low price. 360 Miniature Poses, all different on 4 x 6 clear glossy photo prints. All sent for only $1.00 Post-Paid (No C.O.D.). (Your Money Back If Not Pleased.) COSMO SERVICE, 370 Beach Street, West Haven 16, Conn. DEPT. ZD-PS-7.

LAW...

STUDY AT HOME Legally trained men win higher positions and bigger success in business and public life. Greater opportunities now than ever before. **More Ability: More Prestige: More Money** We guide you step by step. You can train at home during spare time. Degree of LL.B. We furnish all text material, including 14-volume Law Library. Low cost, easy terms. Get our valuable 48-page "Law Training for Leadership" and "Evidence" books FREE. Send NOW.
LASALLE EXTENSION UNIVERSITY, 417 South Dearborn Street
A Correspondence Institution   Dept. L-679   Chicago 5, Ill.

C.G.H. TOMPKINS by Igor Gamow & Scorpio Steele

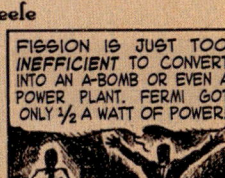

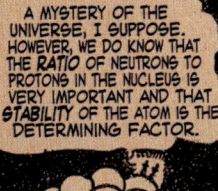

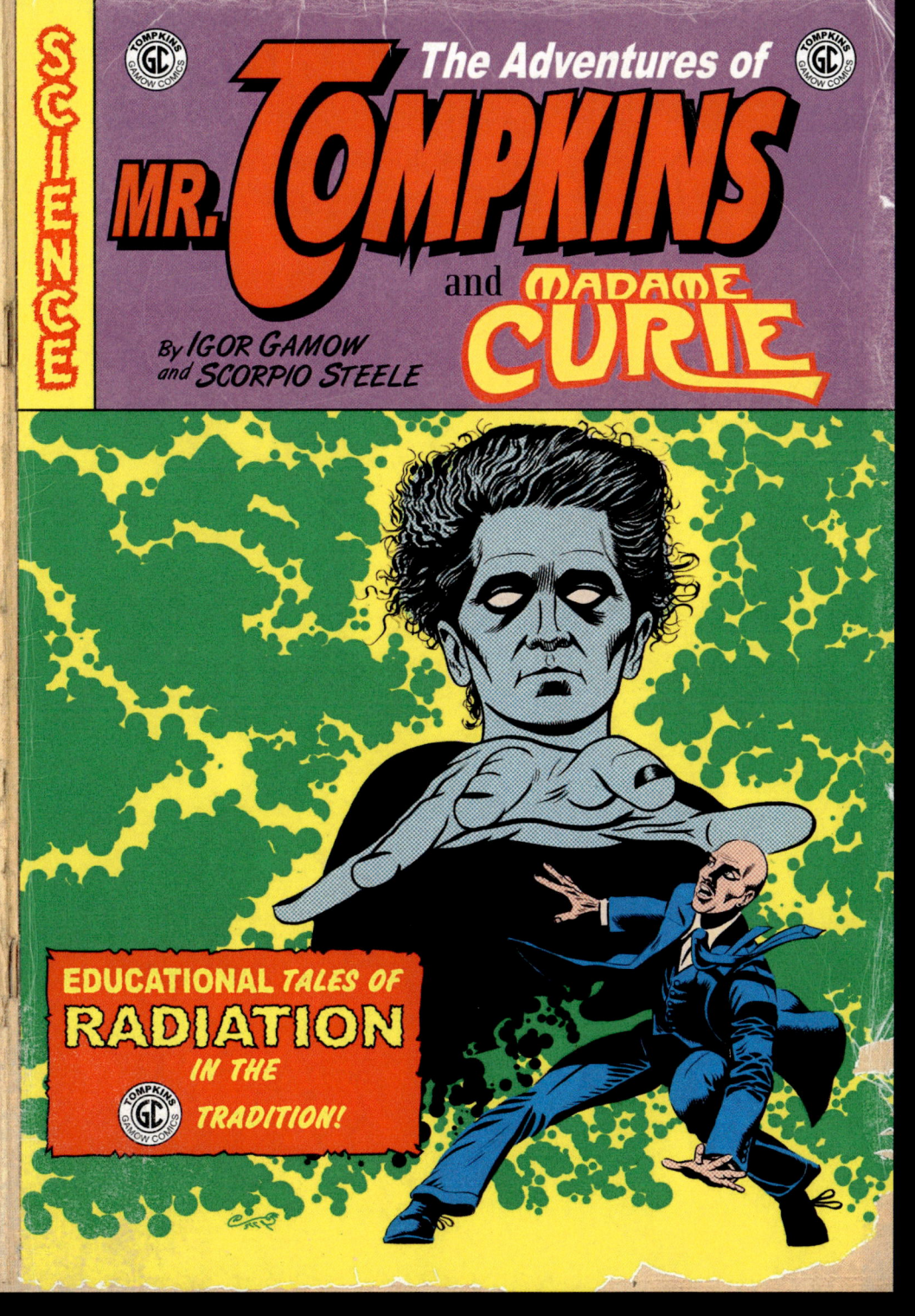

#

"Pierre Curie came to see me and showed a simple and sincere sympathy with my student life. Soon he caught the habit of speaking to me of his dream of an existence consecrated entirely to scientific research, and he asked me to share that life."

Marie Curie

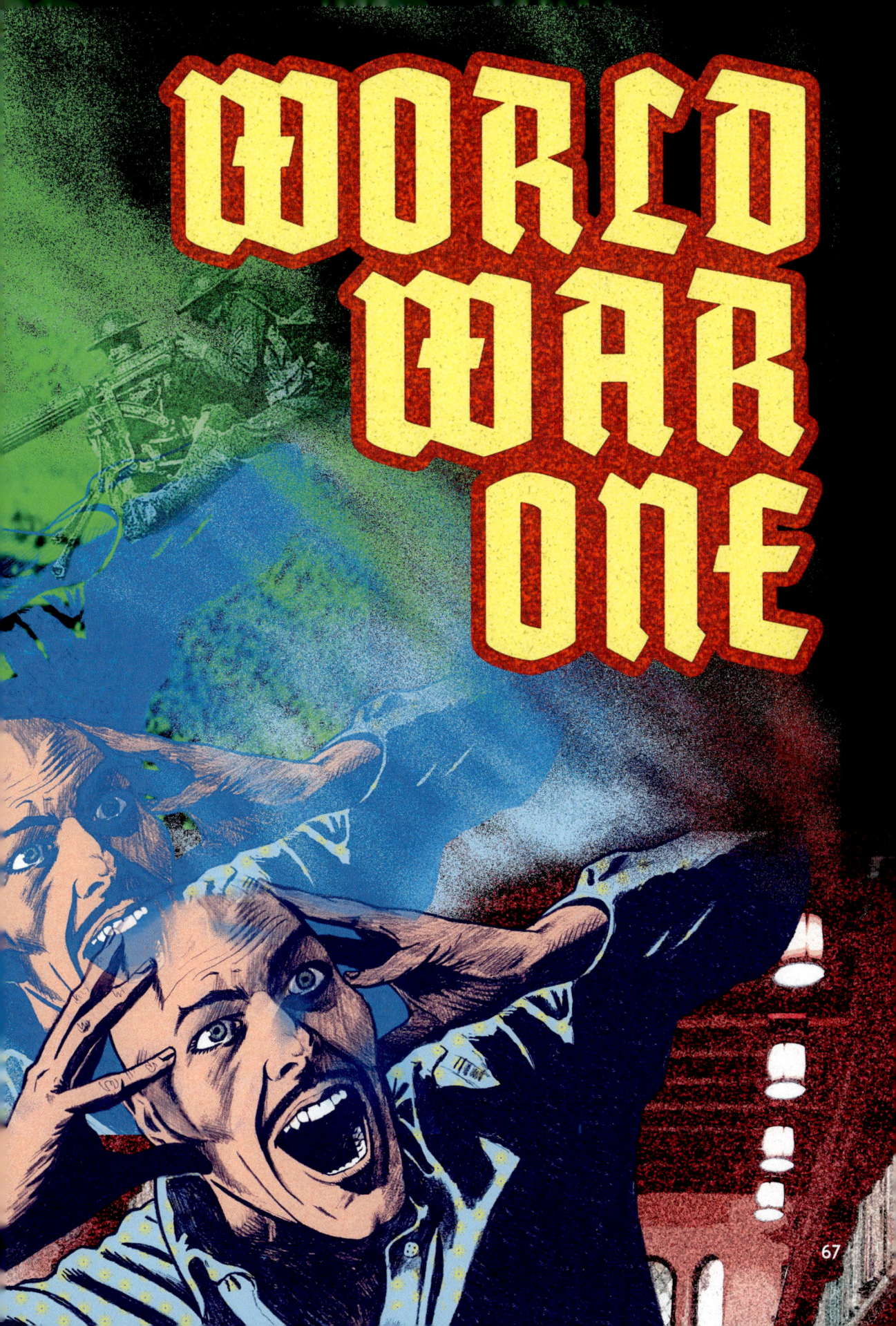

"TO ANSWER YOUR SECOND QUESTION— JUST BEFORE **THE RADIUM INSTITUTE** OPENED, GERMANY INVADED FRANCE AND ALL THE MALE PERSONNEL OF THE INSTITUTE JOINED THE MILITARY TO DEFEND OUR COUNTRY…"

"THE INSTITUTE WAS NOW DESERTED, SO MY SEVENTEEN-YEAR-OLD DAUGHTER, **IRENE**, AND I ALSO VOLUNTEERED, BECOMING BATTLEFIELD NURSE RADIOGRAPHERS."

"TO AID OUR CAUSE WEALTHY AND GENEROUS PARISIANS DONATED TRUCKS-- SUCH AS THE **RENAULT** WE'RE IN AND THE BUNCH YOU SEE HERE--TO CONVERT INTO **MOBILE SURGICAL UNITS** AND TRANSPORTS FOR OUR **X-RAY MACHINES**."

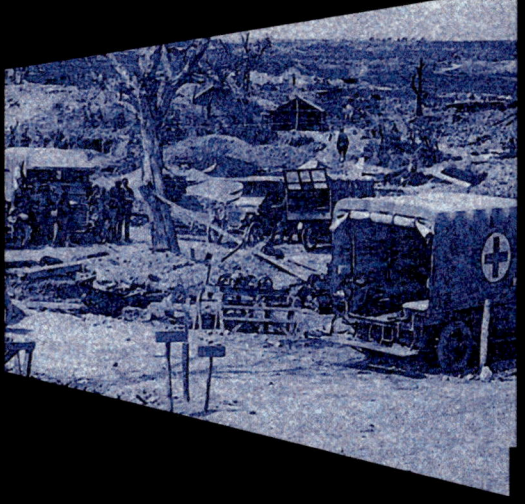

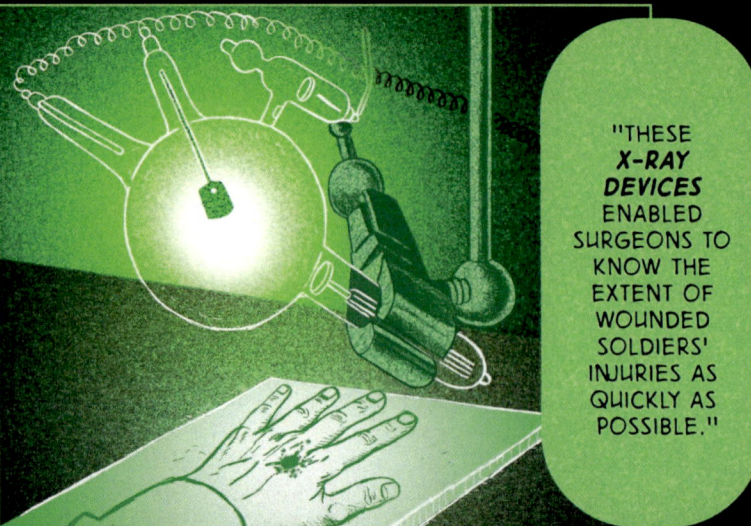
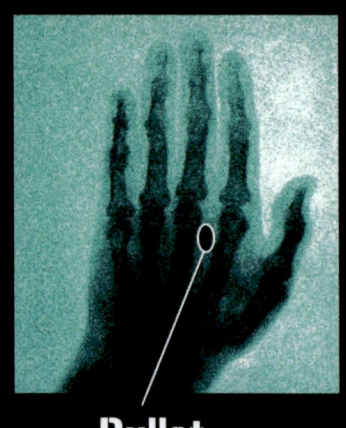

Bullet

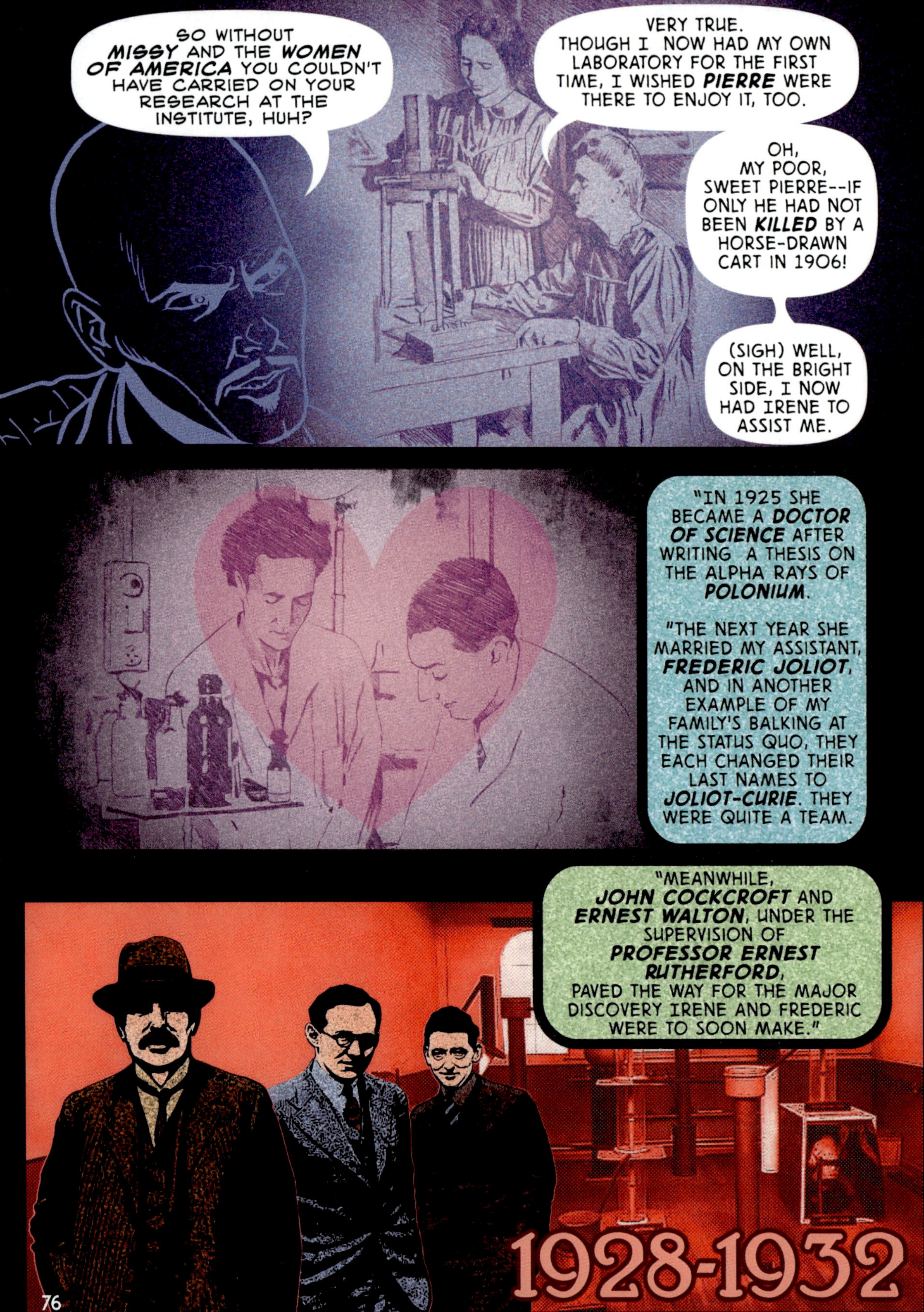

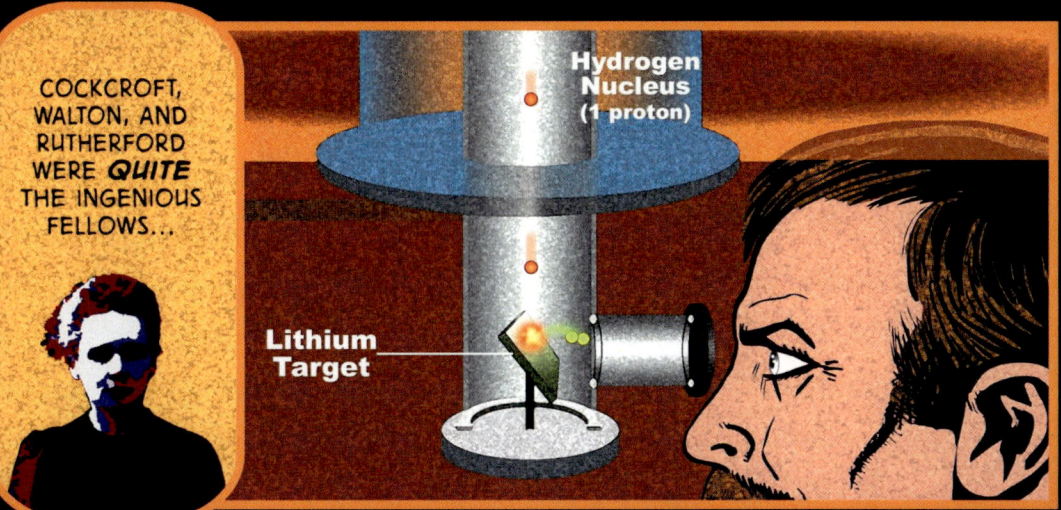

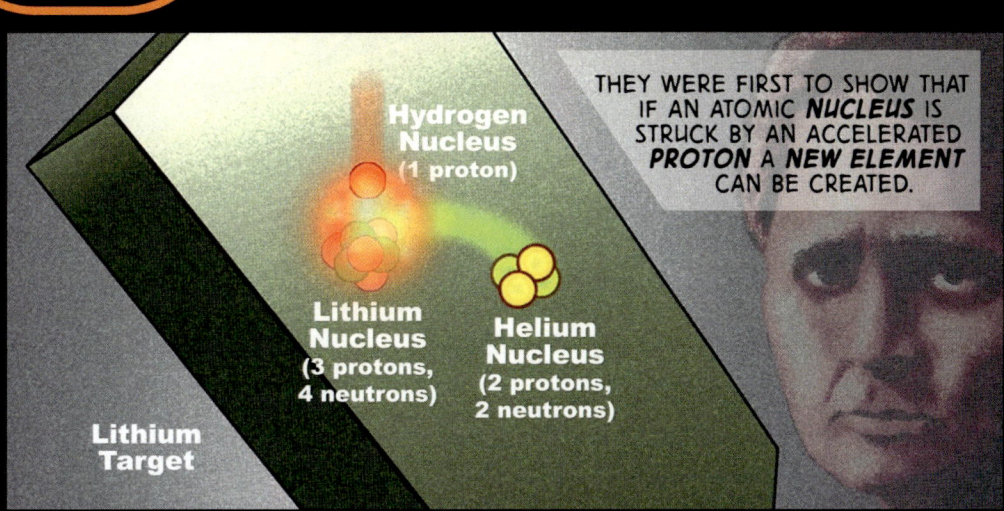

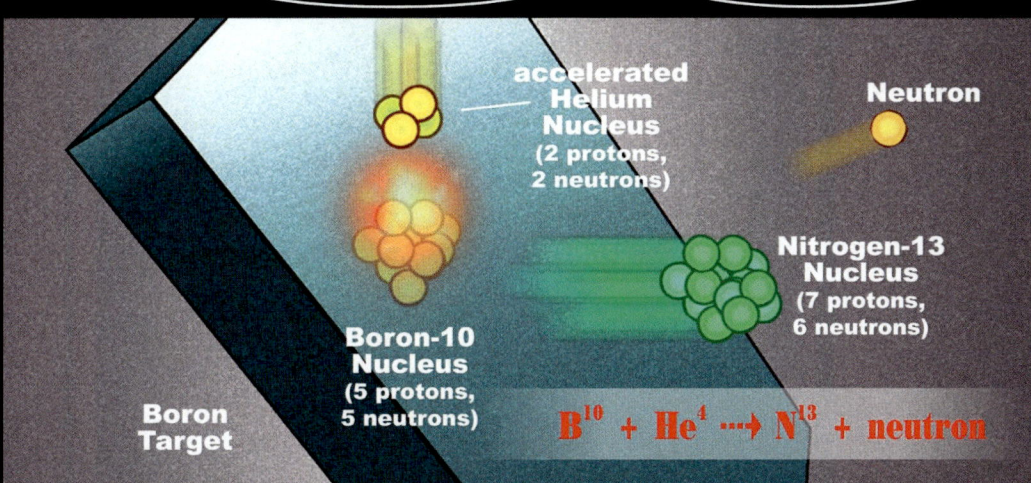

"ISOTOPE? WHAT'S THAT?"

"AH! GOOD QUESTION. AN ISOTOPE IS A VARIANT OF A CHEMICAL ELEMENT THAT HAS A DIFFERENT ATOMIC MASS THAN ITS COUSINS."

"FOR EXAMPLE, **BORON-10** AND **BORON-11** ARE TWO DIFFERENT ISOTOPES OF THE SAME ELEMENT, **BORON**. EACH ISOTOPE HAS THE SAME NUMBER OF PROTONS, (FIVE) BUT BORON-10 HAS FIVE NEUTRONS AND BORON-11 HAS SIX NEUTRONS. THUS, THE NUMBERS OF BOTH THE ISOTOPE AND MASS ARE THE SAME."

"IF, BY ANALOGY, THE **"GAMOW BORON"** IS A TYPE OF AUTOMOBILE, THE **BORON-10** IS A TWO-DOOR **COUPE** WHILE THE **BORON-11** IS A FOUR-DOOR **SEDAN**. EACH IS JUST A DIFFERENT VERSION OF THE SAME CAR."

"IRENE AND FREDERIC SOON REPEATED THE PROCESS WITH ALUMINUM AND MAGNESIUM AND GOT RADIOISOTOPES OF **PHOSPHORUS** FROM THE ALUMINUM AND **SILICON** FROM THE MAGNESIUM."

THEIR WORK RADICALLY TRANSFORMED THE **PERIODIC TABLE OF ELEMENTS** (WHICH QUICKLY GREW TO INCLUDE MORE THAN 400 RADIOISOTOPES) AND EARNED THEM A **NOBEL PRIZE IN CHEMISTRY** IN 1935. ELEVEN YEARS LATER IRENE BECAME THE THIRD DIRECTOR OF THE RADIUM INSTITUTE.

Custom-made radioactive elements have been used for a great variety of products, both industrial and medical.

For example, some of the artificially radioactive elements created are used in common, modern-day **SMOKE DETECTORS**.

piezoelectric sounder (makes alarm noise)

ionization chamber

These "ionization" smoke detectors contain 0.9 **MICROCURIES** of the radioactive element **AMERICIUM-241**, an—

WAITAMINUTE! A MICROCURIE? A SMALL *YOU*?

NO, YOU DUNDERHEADED DANDY!

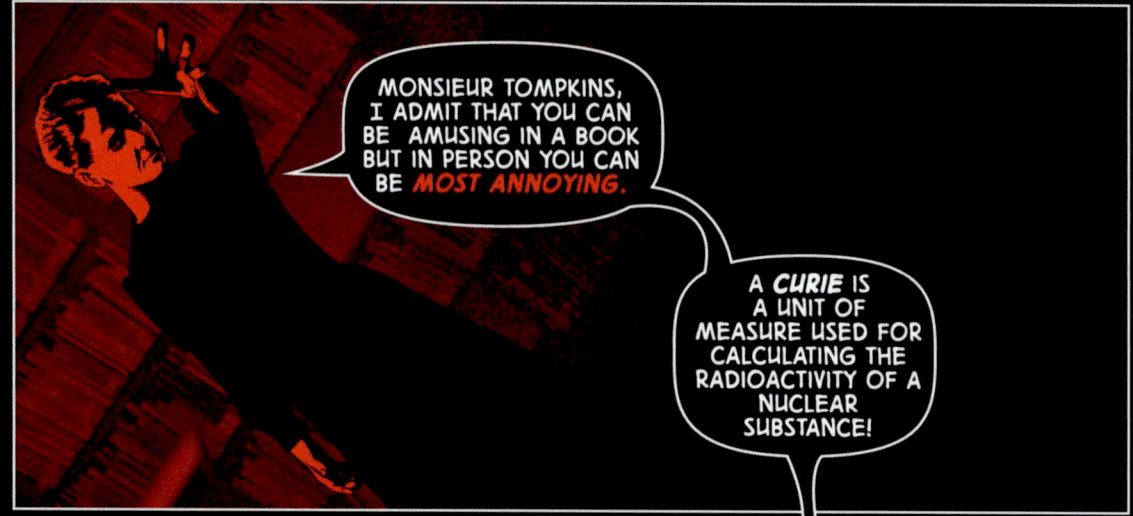

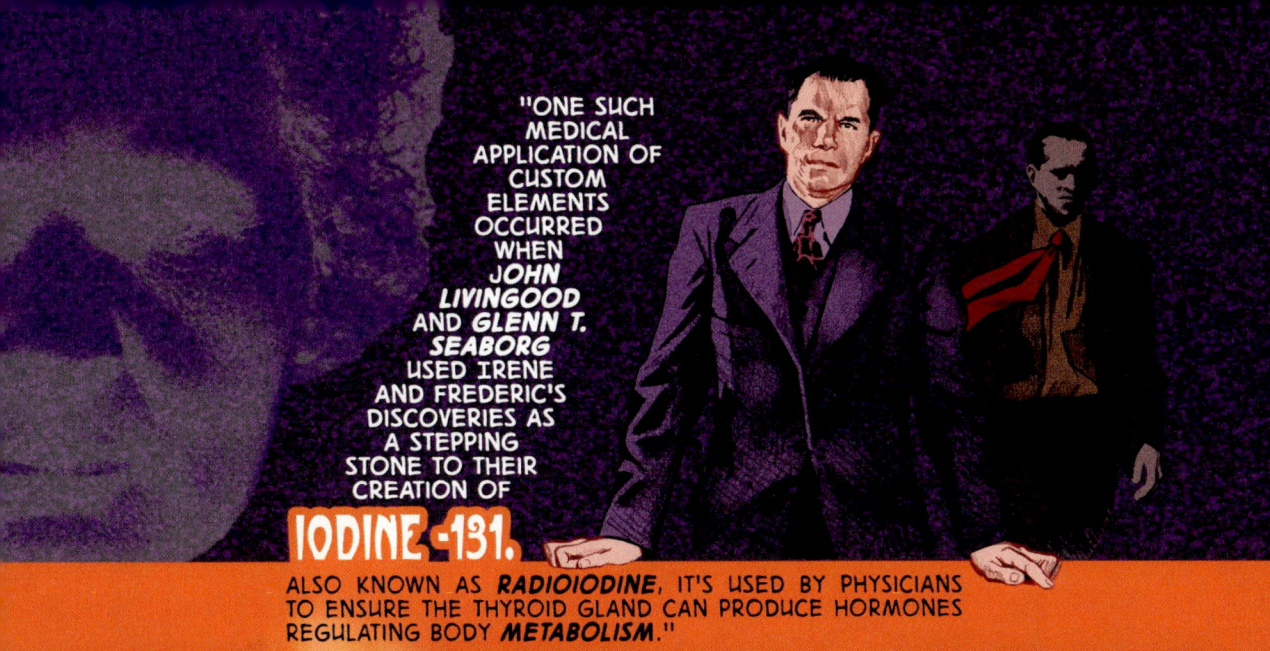

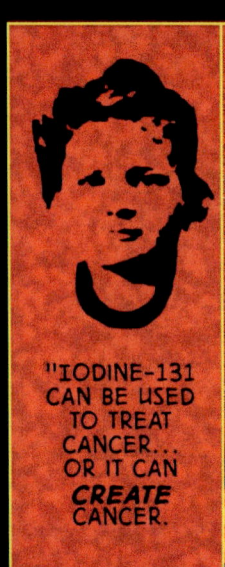

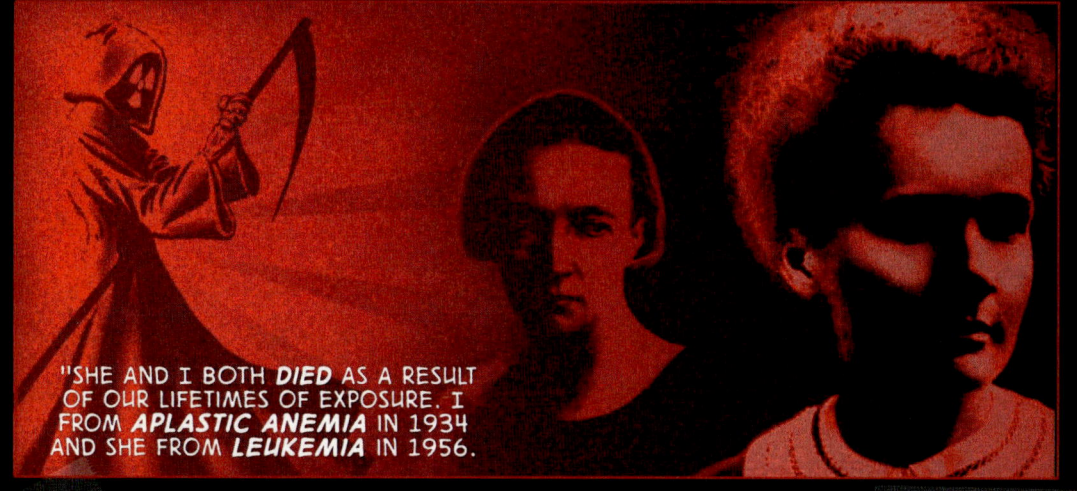

"SHE AND I BOTH **DIED** AS A RESULT OF OUR LIFETIMES OF EXPOSURE. I FROM **APLASTIC ANEMIA** IN 1934 AND SHE FROM **LEUKEMIA** IN 1956."

"IN 1920, FOURTEEN YEARS BEFORE MY **DEATH**, I HAD AN INKLING THAT PERHAPS RADIUM WAS MAKING US SICK-- I HAD **EYESIGHT TROUBLE** AND AN INCESSANT **HUMMING** IN MY EARS.

"UNFORTUNATELY, **PRIDE** PREVENTED ME FROM LISTENING TO MY **INTUITION**. I JUST PRETENDED THAT NOTHING WAS THE MATTER, CONFIDING ONLY TO MY **SISTER** THAT ANYTHING WAS WRONG.

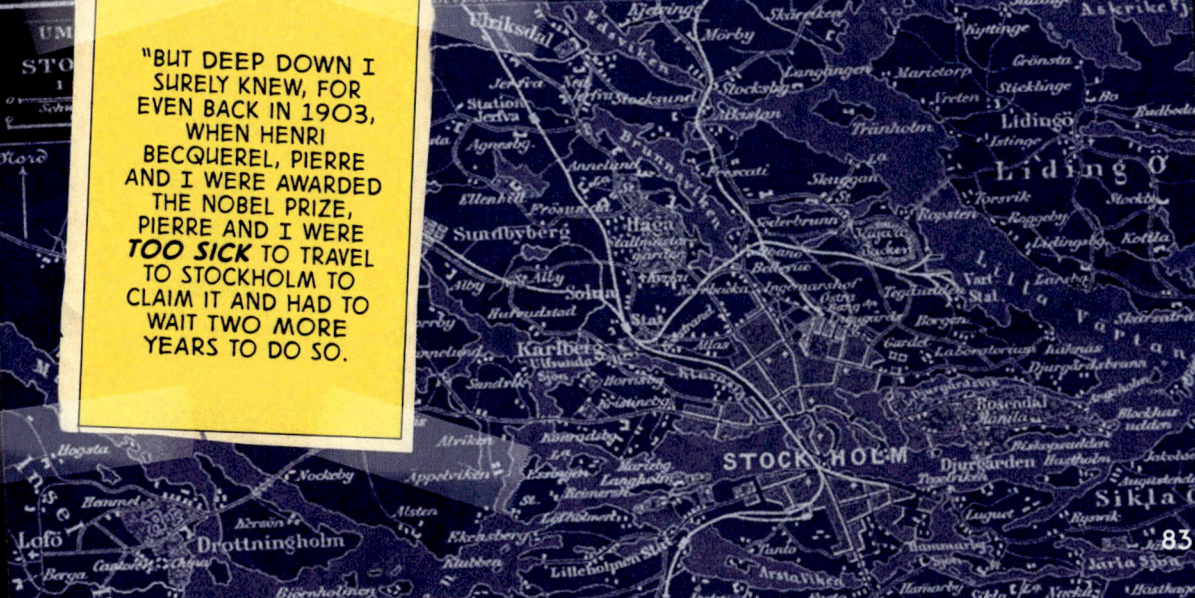

"BUT DEEP DOWN I SURELY KNEW, FOR EVEN BACK IN 1903, WHEN HENRI BECQUEREL, PIERRE AND I WERE AWARDED THE NOBEL PRIZE, PIERRE AND I WERE **TOO SICK** TO TRAVEL TO STOCKHOLM TO CLAIM IT AND HAD TO WAIT TWO MORE YEARS TO DO SO.

"BE THIS AS IT MAY, PIERRE AND I **WERE** AWARE OF THE POWER OF RADIOACTIVITY TO **HEAL**, IRONICALLY DUE TO PIERRE EXPERIMENTING UPON HIMSELF.

"FOR TEN HOURS HE STRAPPED A VIAL OF RADIUM SALT TO HIS OWN FOREARM...

 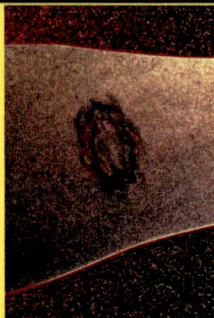 

"...THEN NOTED THE RESULTING BURNS, SCABS AND DEAD TISSUE.

"AS RADIATION KILLED THE TISSUE, PIERRE WONDERED IF IT WOULD KILL **TUMORS**.

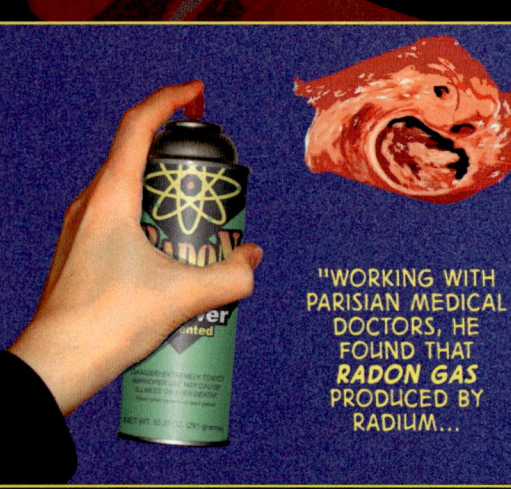 

"WORKING WITH PARISIAN MEDICAL DOCTORS, HE FOUND THAT **RADON GAS** PRODUCED BY RADIUM...

"...WHEN PLACED CLOSE TO A GROWING TUMOR...

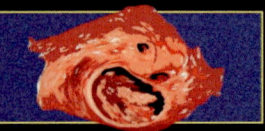 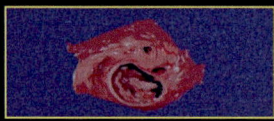  

"...**KILLED** GROWING CANCER CELLS RATHER **SELECTIVELY**."

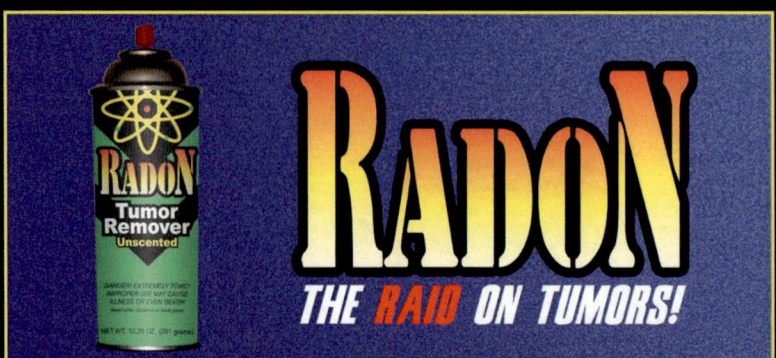

RADON — THE RAID ON TUMORS!

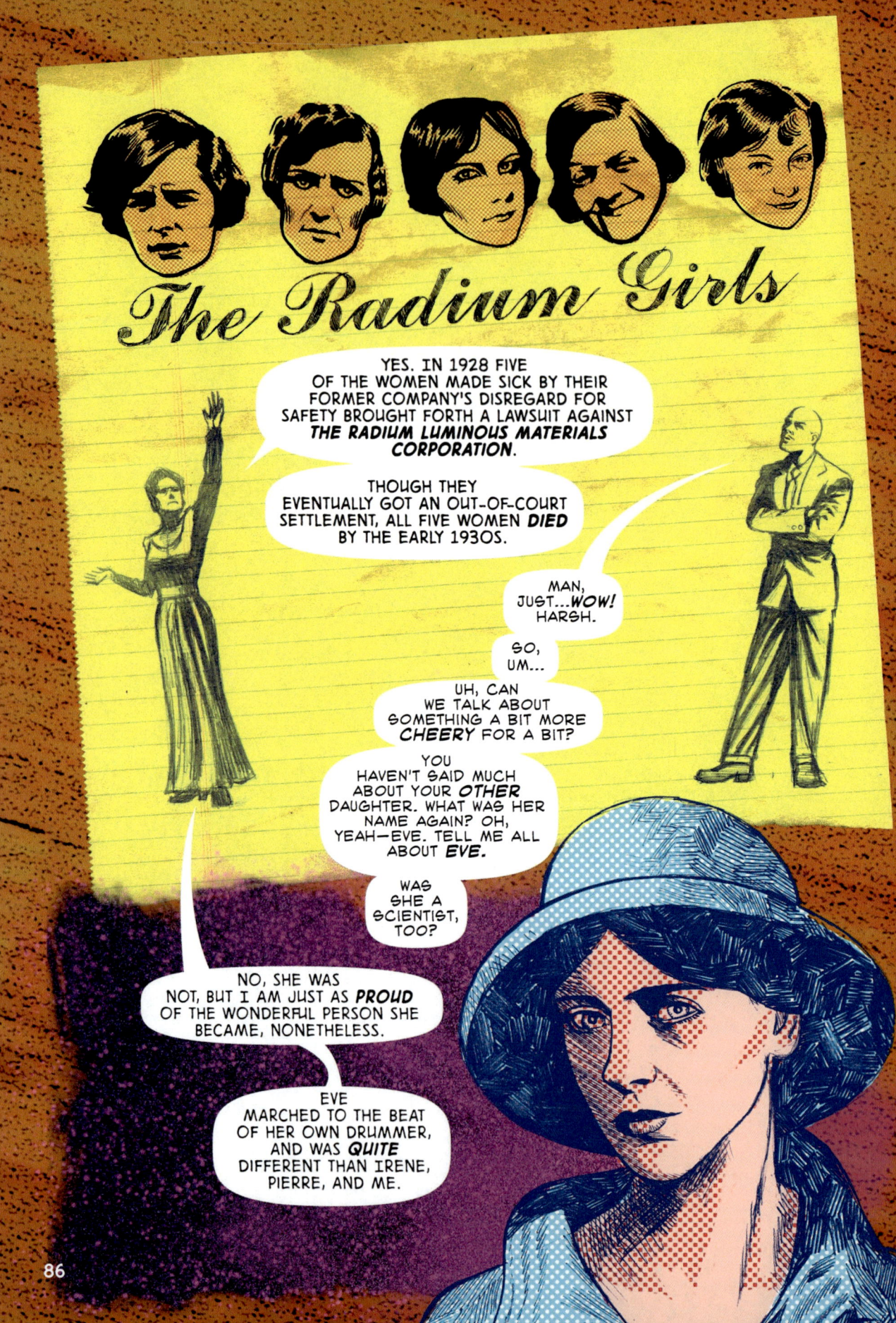

TOMPKINS INTERNATIONAL presents
IN **SCORPIOCOLOR** AND **GAMOW·VISION**

EVE CURIE
"Eve"
THE GIRL WITH THE RADIUM EYES

OUR FAMILY WAS BASICALLY SHY AND MODEST, AND DID NOT SEEK ATTENTION. **EVE**, THOUGH, WAS **OUTGOING**, PROUDLY WORE **JEWELRY** AND **MAKE-UP.** ALMOST ALWAYS SHE WAS THE **CENTER OF ATTENTION**.

WHEN SHE, IRENE AND I TOURED AMERICA IN 1921, EVE WAS THE MAIN ONE WHO DEALT WITH THE NEWS MEDIA, WHICH DUBBED HER *"THE GIRL WITH THE RADIUM EYES."*

YEARS LATER EVE AND HER HUSBAND, **HENRI LABOUISSE**, WERE TWO OF THE EARLY FOUNDERS OF WHAT BECAME KNOWN AS **unicef**

HE WAS AMERICAN AMBASSADOR TO GREECE AND UNICEF'S EXECUTIVE DIRECTOR WHEN IT WON THE NOBEL PEACE PRIZE IN 1965.

SHE HERSELF WAS DIRECTOR OF UNICEF IN GREECE FROM 1962 TO 1965.

SO, IN A **BROAD** SENSE THE CURIES ARE ASSOCIATED WITH **FOUR** NOBEL PRIZES.

OUI, MONSIEUR, BUT... I DO NOT LIKE TO **BRAG**.

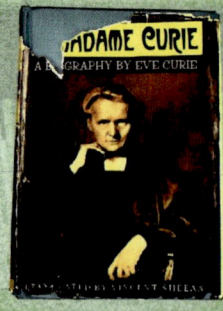

SCIENCE

The Adventures of MR. TOMPKINS and DARWIN

By IGOR GAMOW *and* SCORPIO STEELE

EDUCATIONAL *TALES OF* EVOLUTION *IN THE* TRADITION!

THERE HE IS!

HE'S WALKING ON STAGE!

GOOD EVENING, LADIES AND GENTLEMEN. THANK YOU FOR JOINING ME!

I AM PROFESSOR IGOR, AND TONIGHT'S LECTURE IS ENTITLED...

THE ORIGIN OF "ON THE ORIGIN OF SPECIES"!

To attend Professor Igor's lecture, please visit the Mr. Tompkins page of www.gamow.com

Hours later, after Tompkins returned home...

...he dreamed...

"As dogs, cats, horses, and probably all the higher animals, even birds, as is stated on good authority, have vivid dreams, and this is shewn by their movements and voice, we must admit that they possess some power of imagination…"

Charles Darwin
The Descent of Man

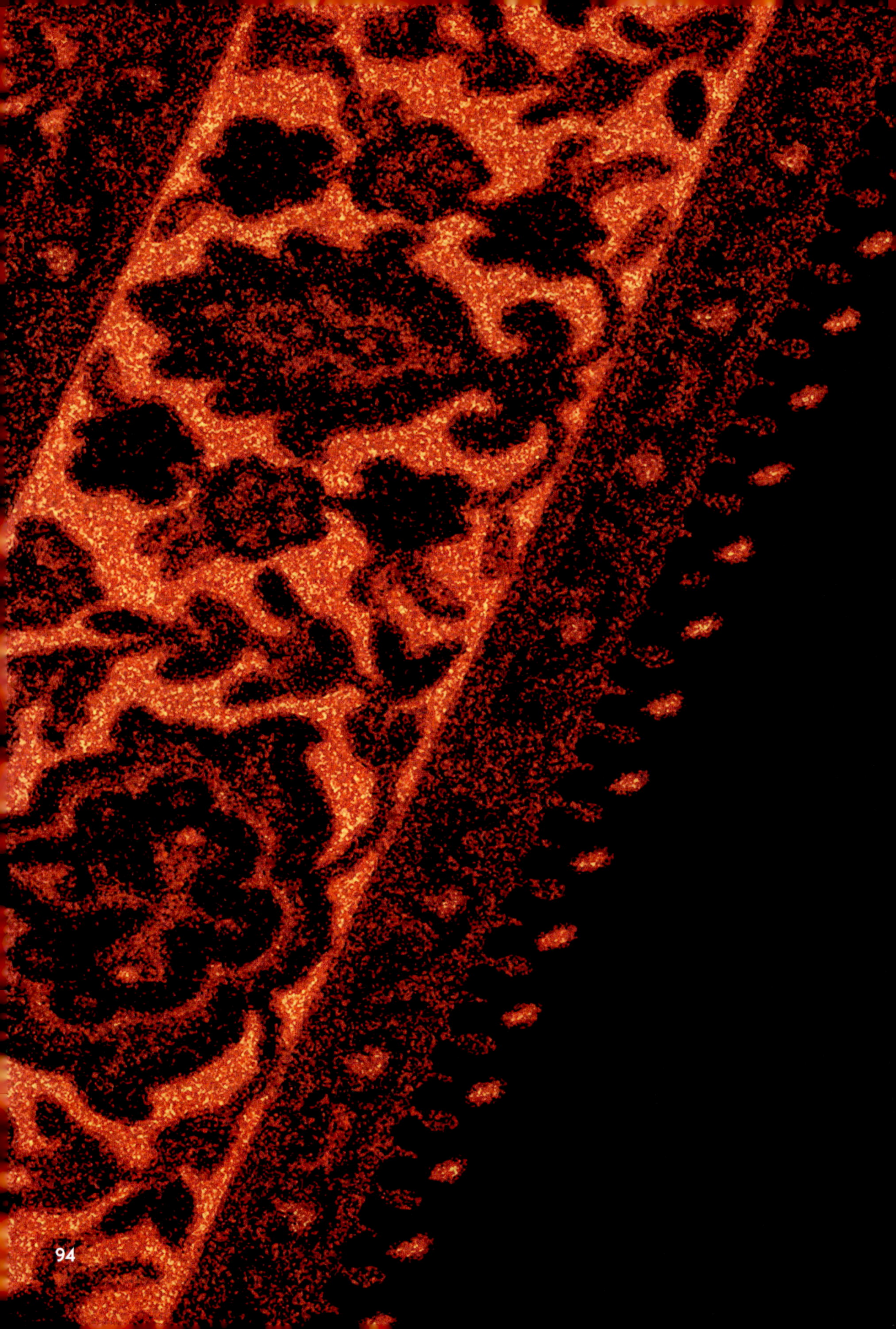

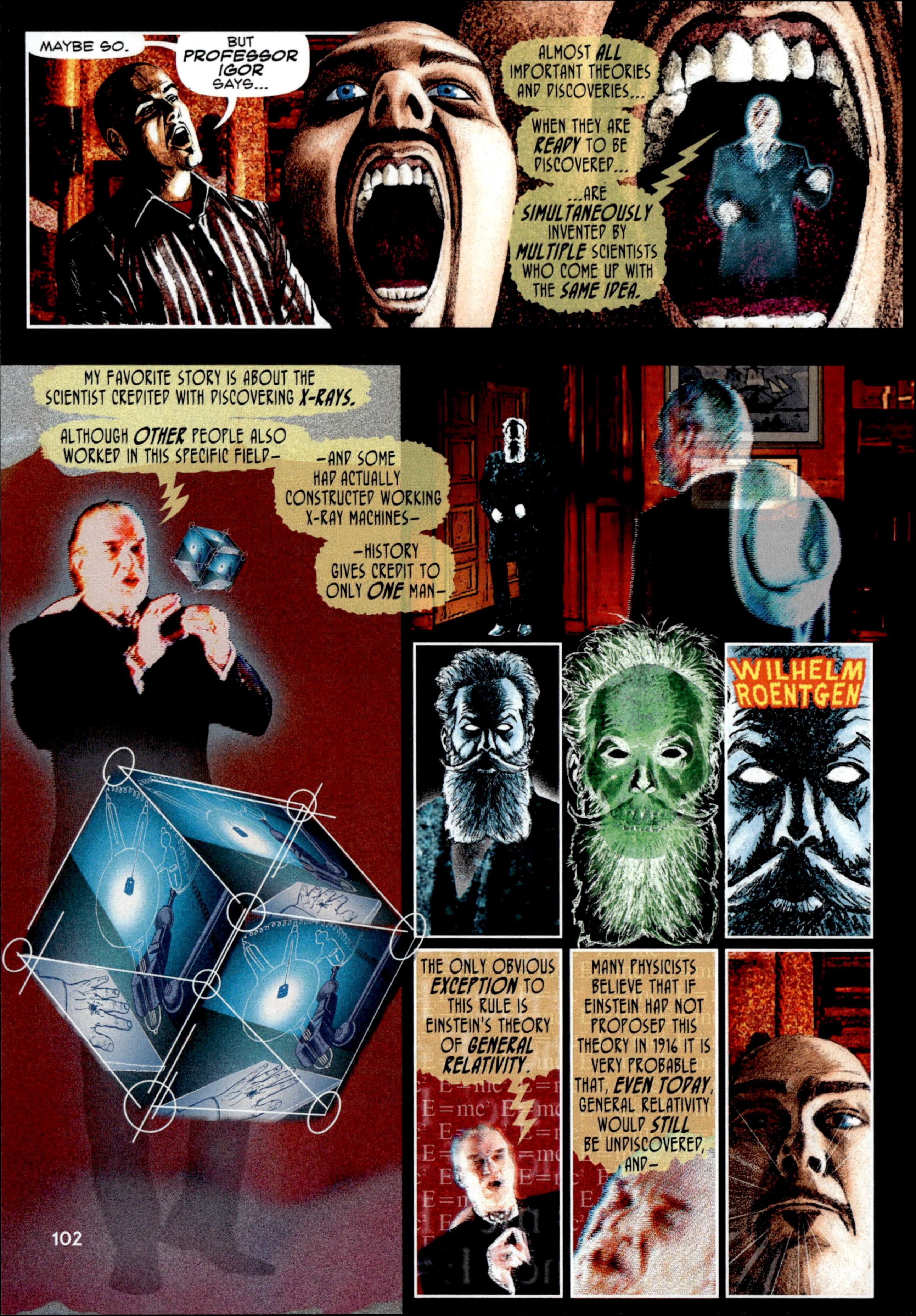

MAYBE SO. BUT PROFESSOR IGOR SAYS...

ALMOST *ALL* IMPORTANT THEORIES AND DISCOVERIES... WHEN THEY ARE *READY* TO BE DISCOVERED... ...ARE *SIMULTANEOUSLY* INVENTED BY *MULTIPLE* SCIENTISTS WHO COME UP WITH THE *SAME IDEA*.

MY FAVORITE STORY IS ABOUT THE SCIENTIST CREDITED WITH DISCOVERING *X-RAYS*.

ALTHOUGH *OTHER* PEOPLE ALSO WORKED IN THIS SPECIFIC FIELD—

—AND SOME HAD ACTUALLY CONSTRUCTED WORKING X-RAY MACHINES—

—HISTORY GIVES CREDIT TO ONLY *ONE* MAN—

WILHELM ROENTGEN

THE ONLY OBVIOUS *EXCEPTION* TO THIS RULE IS EINSTEIN'S THEORY OF *GENERAL RELATIVITY*.

MANY PHYSICISTS BELIEVE THAT IF EINSTEIN HAD NOT PROPOSED THIS THEORY IN 1916 IT IS VERY PROBABLE THAT, *EVEN TODAY*, GENERAL RELATIVITY WOULD *STILL* BE UNDISCOVERED, AND—

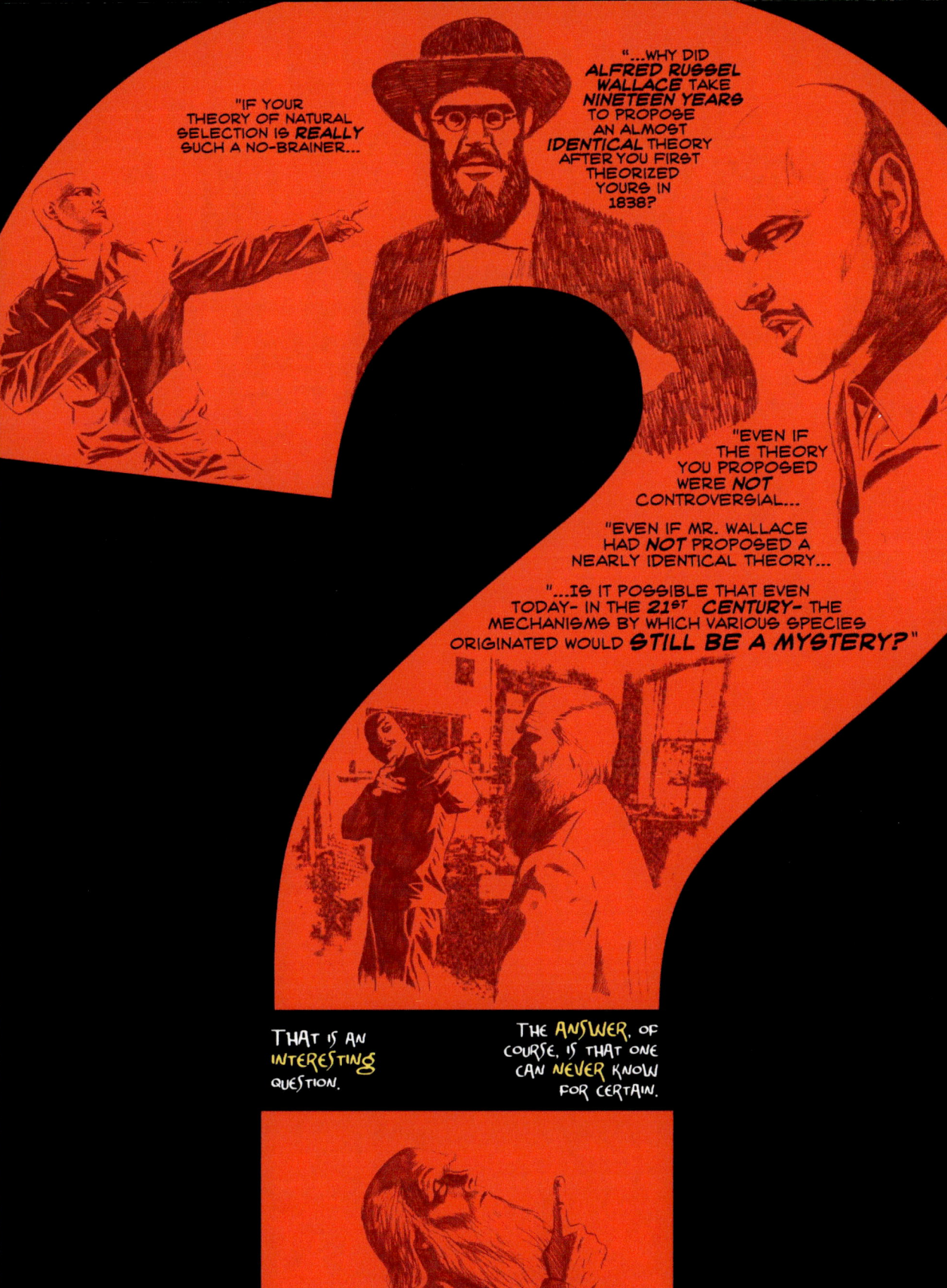

THERE USUALLY ARE SOME...

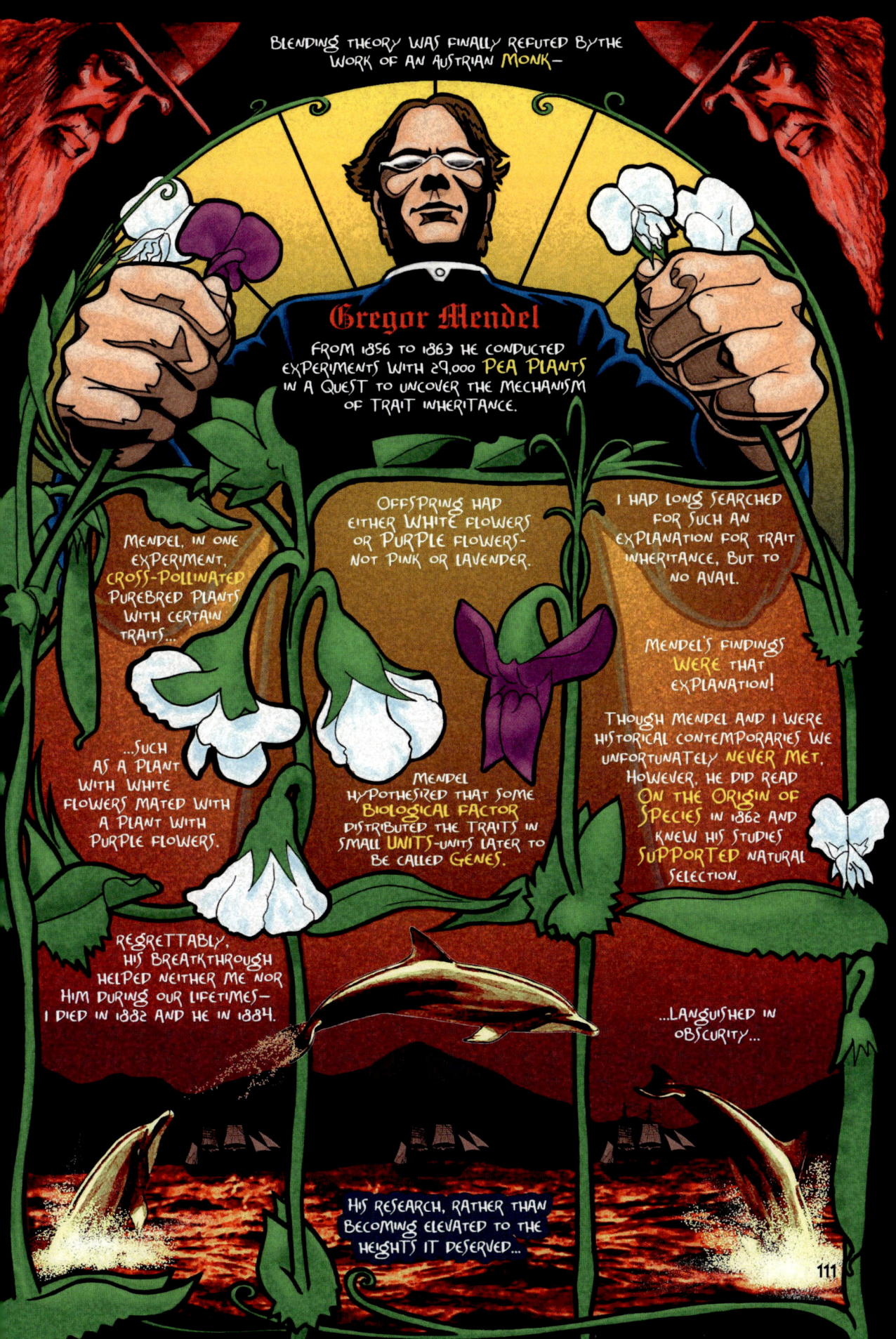

112

...UNTIL 1900, THAT IS, WHEN OTHER SCIENTISTS CAME TO THE SAME CONCLUSIONS AS MENDEL...

...ONLY TO LEARN THAT MENDEL HAD BEEN THERE FIRST!

WELL, I'M GLAD MENDEL'S RESEARCH EVENTUALLY **SUPPORTED** NATURAL SELECTION..

BUT AFTER ALL THIS I'M, UH... **STILL** A BIT CONFUSED AS TO WHAT NATURAL SELECTION **ACTUALLY IS**.

AH, MY APOLOGIES, SIR. **NATURAL SELECTION** IS THE PRINCIPLE BY WHICH EACH SLIGHT VARIATION OF AN ORGANISM'S **TRAIT**, IF **USEFUL**, IS PRESERVED.

OR— HOW A SPECIES OF ORGANISM, IN ITS STRUGGLE FOR EXISTENCE, HEREDITARILY PASSES ON **ADVANTAGEOUS TRAITS** TO ITS OFFSPRING.

I MUST ADD THAT MY THEORY PRESUPPOSES THAT MOTHER NATURE PRODUCES **MORE ORGANISMS** THAN AN ENVIRONMENT CAN SUPPORT...

...THUS INCREASING THE SPECIES' CHANCES FOR SURVIVAL.

BY **JOVE**, I THINK I'VE GOT IT! **THANKS!**

SO, THEN— I SUPPOSE YOU **CAME UP** WITH NATURAL SELECTION **DURING** YOUR FIVE YEAR MISSION ABOARD **THE BEAGLE?**

ACTUALLY... **NO!** BUT YOUR QUESTION REFLECTS A COMMON MISCONCEPTION THAT AROSE SOON AFTER I PUBLISHED **ON THE ORIGIN OF SPECIES** IN 1859. THE THEORY OF NATURAL SELECTION CAME TO ME IN 1838— **TWO YEARS** AFTER THE END OF MY VOYAGE.

THE TREK **DID**, HOWEVER, LAY THE GROUND WORK FOR ME TO EVENTUALLY CONSTRUCT MY THEORY.

"GROUND WORK."

THAT'S AN IRONIC TERM TO DESCRIBE A **SEA VOYAGE**.

BUT... HOW'D YOU DECIDE TO EVEN **TAKE** THE TRIP?

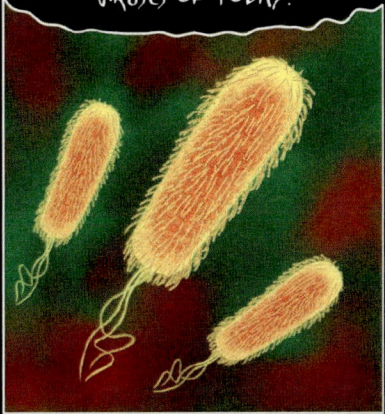

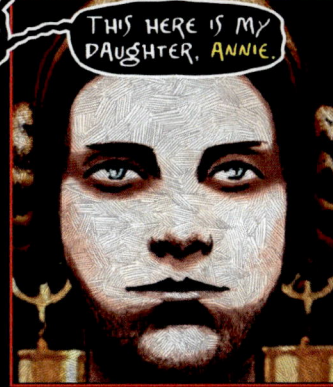

...he dreamed...

"It requires indeed some courage to undertake a labor of such far-reaching extent; this appears, however, to be the only right way by which we can finally reach the solution of a question the importance of which cannot be overestimated in connection with the history of the evolution of organic forms."

Gregor Mendel

"EDELWEIIISSSSS..."
"EDELWEIIISSSSS..."
"EVERY MORNING YOU GREEEET MEEE..."
GAHH!

MEIN GOTT! HERR TOMPKINS, I DO DECLARE THAT YOU DID STARTLE THIS POOR MONK OUT OF HIS VERY WITS!
SORRY ABOUT THAT, BROTHER MENDEL.
I GOT CARRIED AWAY FLYING! NOTHING LIKE A GOOD DREAM OF FLIGHT, EH?
BUT, HEY— I SUPPOSE THAT SINCE YOU KNOW MY NAME... MY REPUTATION MUST PRECEDE ME?
OH, YES INDEED! I AM QUITE A FAN OF YOURS, GOOD SIR!
IN FACT, IF YOU DON'T MIND...
...I WOULD SURELY APPRECIATE YOUR AUTOGRAPH UPON THIS FOR ME!
BUT OF COURSE!

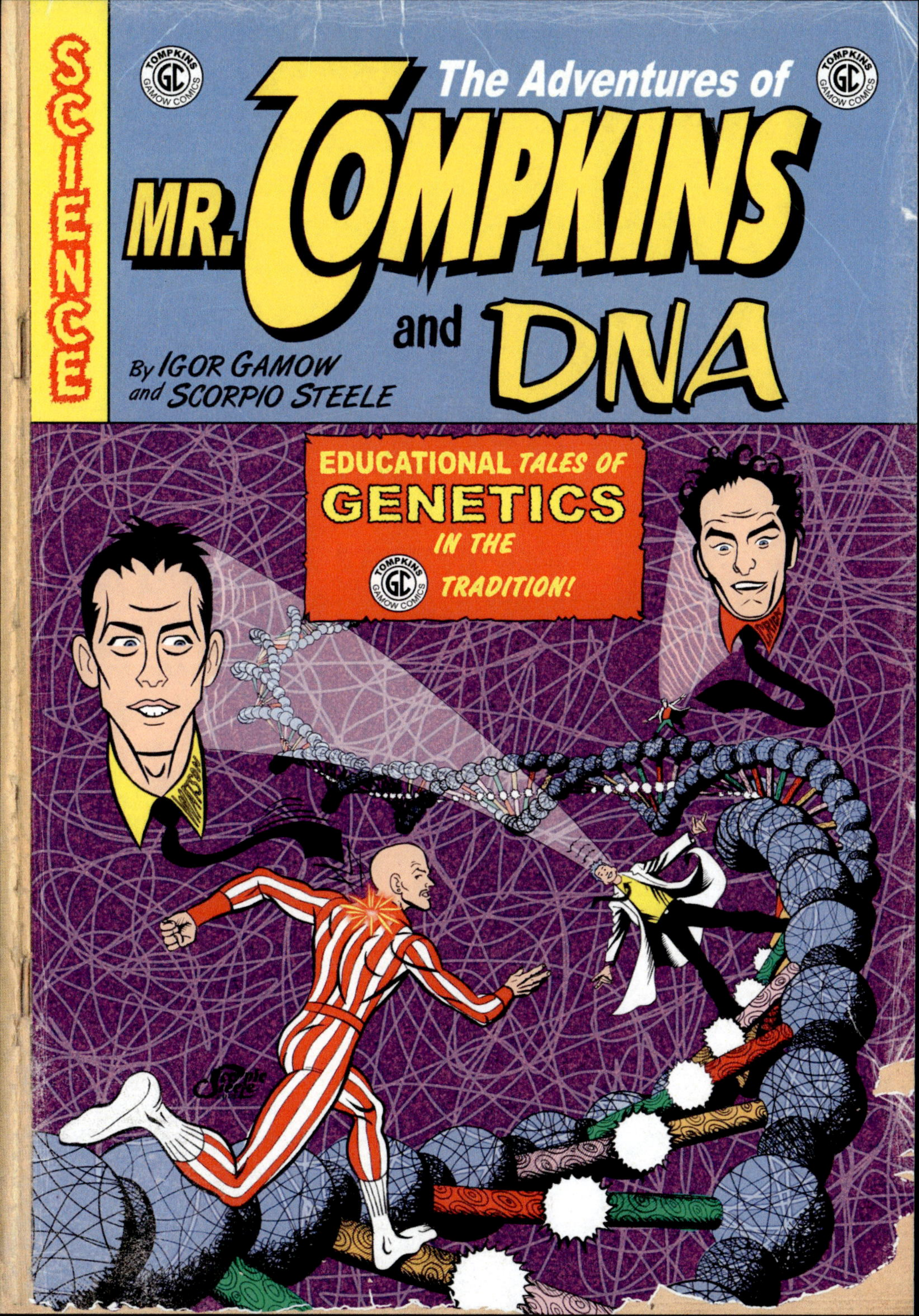

"If you succeed with your first dream, it helps. You know, people trust you, possibly, for the second one. They give you a chance to play out your second one."

James D. Watson

"I stand amid the roar
Of a surf-tormented shore..."

"And I hold within my hand
Grains of the golden sand..."

"How few! yet how they creep
Through my fingers to the deep..."

"While I weep —
while I weep!"

"O God! can I not grasp
Them with a tighter clasp?"

"O God! can I not save
One from the pitiless wave?"

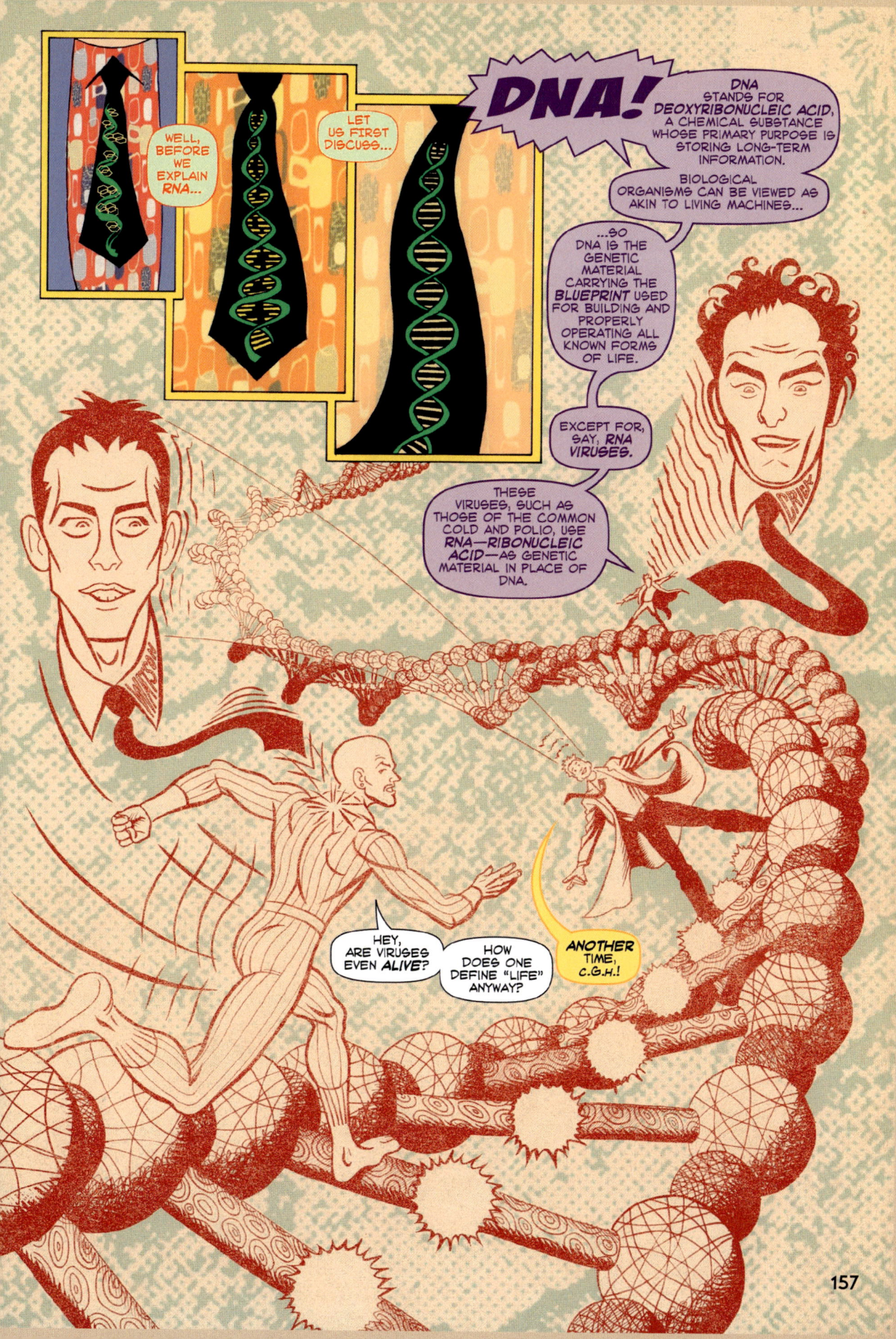

"JUST A FEW NIGHTS AGO I LEARNED ABOUT GENES FROM *MENDEL* HIMSELF!"

"COOL, C.G.H.!"

"SO, WHAT YOU SEE IS A STYLIZED STRUCTURAL DNA MODEL INSPIRED BY THE ONE FRANCIS AND I DISCOVERED BACK IN 1953."

"KEEP THIS IN MIND—YOU WOULD *NOT* ACTUALLY SEE THIS THROUGH AN ELECTRON MICROSCOPE."

"THE "RAILS" OF THE SPIRAL LADDER—ITS "BACKBONES"—ARE MADE OF *PHOSPHATE* GROUPS BONDED TO FIVE-ATOM *DEOXYRIBOSE SUGAR* MOLECULES."

"ATTACHED TO EACH DEOXYRIBOSE MOLECULE IS ANOTHER MOLECULE CALLED A *NUCLEOBASE*. BASES COME IN FOUR "FLAVORS": *ADENINE, CYTOSINE, GUANINE* AND *THYMINE*."

"EACH BASE IS VERY PICKY ABOUT WHICH OTHER BASE IT'LL "HOOK UP"—ADENINE PAIRS ONLY WITH THYMINE AND CYTOSINE WITH GUANINE. A *HYDROGEN BOND* HOLDS THEM TOGETHER LIKE A NEWBORN INFANT."

"THE PHOSPHATE-SUGAR-BASE TRIO IS CALLED A *NUCLEOTIDE*. THE NUCLEOTIDE SEQUENCE ALONG THE BACKBONES IS A CODED GENETIC "TEXT" SPECIFYING THE AMINO ACID SEQUENCE WITHIN PROTEINS."

"EACH SERIES OF THREE ADJACENT BASES IN A POLYNUCLEOTIDE LINK OF DNA OR RNA IS CALLED A *CODON* BECAUSE IT CODES FOR ONE SPECIFIC AMINO ACID."

"GOOD OL' *GEORGE GAMOW* WAS THE FIRST TO POSTULATE THE CODON'S TRIPLICATE NATURE!"

"DURING A PROCESS CALLED *TRANSCRIPTION* THE CODED TEXT IS READ BY COPYING—*TRANSCRIBING*—DNA SEGMENTS INTO THE RELATED NUCLEIC ACID *RNA*."

"RNA IS SIMILAR TO DNA IN THAT IT'S ALSO MADE OF NUCLEOTIDES, THOUGH INSTEAD OF DNA'S DEOXYRIBOSE SUGAR IT USES *RIBOSE*, WHICH HAS ONE MORE OXYGEN ATOM THAN DNA'S SUGAR."

"ALSO, INSTEAD OF THYMINE, RNA USES *URACIL* TO BOND WITH ADENINE."

"ANYHOO! DNA AND RNA—ALONG WITH PROTEINS—ARE THREE *MACROMOLECULES* INCLUDED IN ALL FORMS OF LIFE."

"DNA CODES FOR RNA AND PROTEINS, TOO, AND *GENES* ARE WHAT WE CALL THE DNA SEGMENTS OF CODED GENETIC INFO."

158

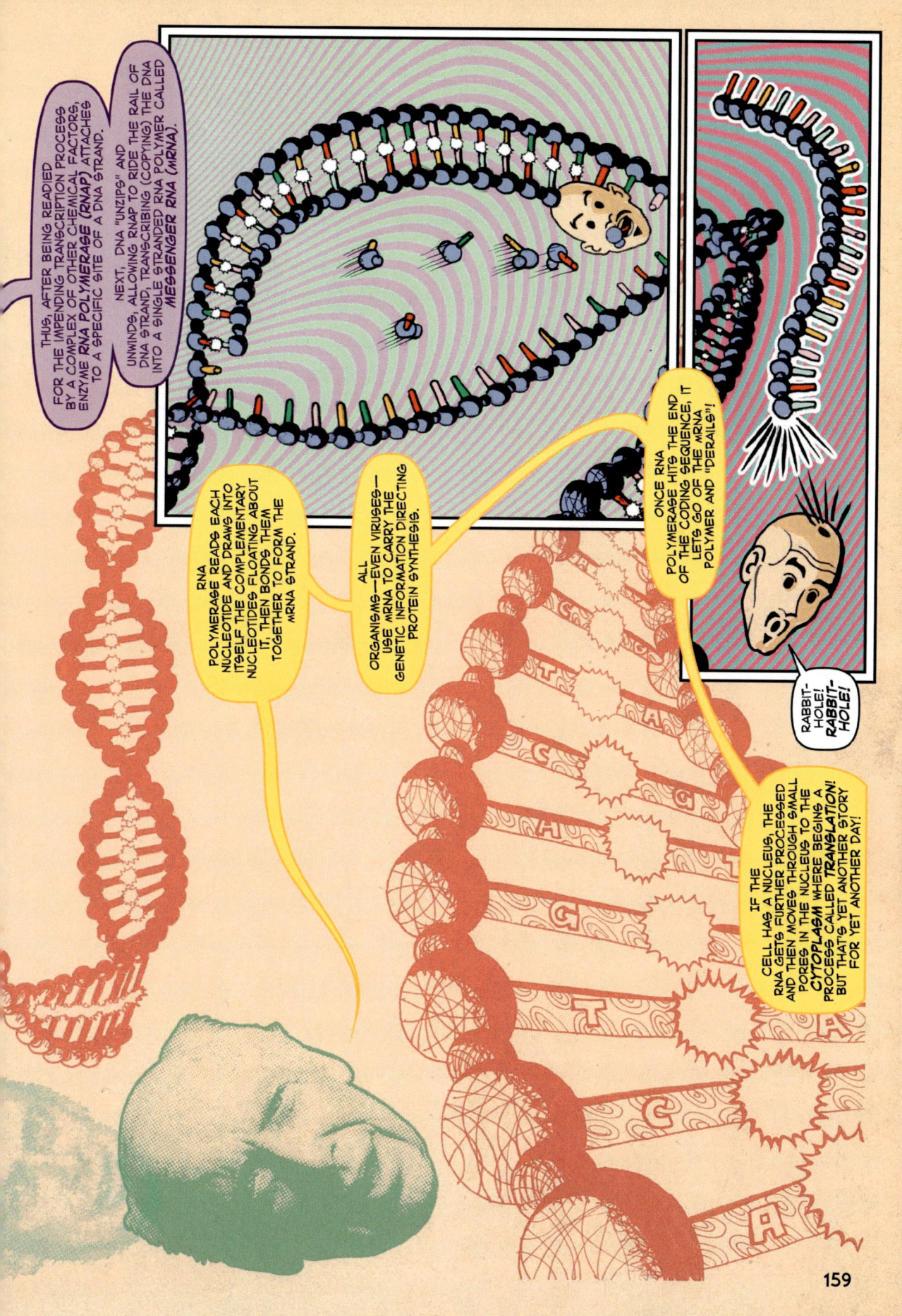

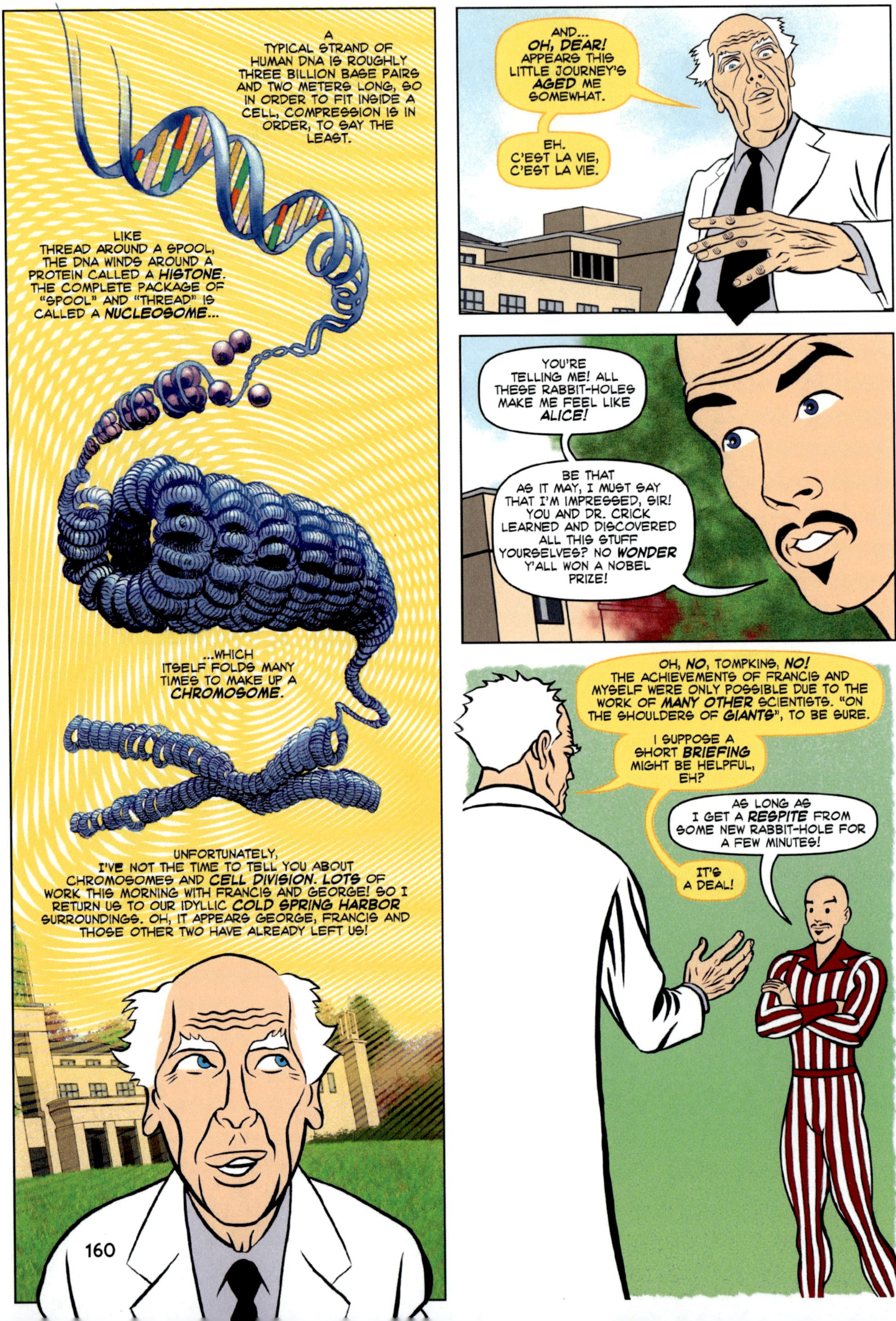

FRIEDRICH MIESCHER

IN 1869, WHILE STUDYING THE CHEMISTRY OF THE CELL NUCLEUS USING **LEUKOCYTES***, SWISS DOCTOR AND BIOLOGIST **FRIEDRICH MIESCHER** PERSUADED A LOCAL SWISS HOSPITAL TO GIVE HIM PUS-COVERED BANDAGES (PUS IS RICH IN LEUKOCYTES). HE PLANNED TO JUST WASH THE BANDAGES, FILTER OUT THE LEUKOCYTES AND SEE WHAT PROTEINS MADE UP THE CELLS.

INSTEAD, HE NOTICED WITHIN THE NUCLEUS A **STRANGE MICROSCOPIC SUBSTANCE!** DIFFERENT FROM ANY PROTEIN, IT CONTAINED MORE PHOSPHOROUS AND WAS RESISTANT TO PROTEIN DIGESTION. HE CALLED THE SUBSTANCE **NUCLEIN**, WHICH WAS LATER RENAMED **NUCLEIC ACID**.

PHOEBUS LEVENE

IN THE EARLY 20TH CENTURY, RUSSIAN-AMERICAN BIOCHEMIST **PHOEBUS LEVENE** DETERMINED THE DIFFERENCES BETWEEN THE TWO KINDS OF NUCLEIC ACID. HE ALSO DISCOVERED AND NAMED BOTH **RIBOSE** AND **DEOXYRIBOSE**.

FURTHER, HE ALSO DEMONSTRATED THAT NUCLEIC ACIDS ARE BUILT OF NUCLEOTIDES AND NUCLEOBASES AND EVEN NAMED THESE BUILDING BLOCKS!

FREDERICK GRIFFITH

MICROBIOLOGIST **FREDERICK GRIFFITH** DEMONSTRATED IN 1928 THAT A BACTERIUM CAN MODIFY ITS FORM AND FUNCTION WHEN HE TRANSFERRED TRAITS OF THE "SMOOTH" FORM OF **PNEUMOCOCCUS** BACTERIA TO ITS "ROUGH" FORM BY MIXING DEAD "SMOOTH" WITH THE LIVE "ROUGH".

UNTIL THIS POINT, A BACTERIUM'S FORM AND FUNCTION WERE CONSIDERED IMMUTABLE, SO GRIFFITH ATTRIBUTED THE CHANGES TO SOME UNKNOWN **TRANSFORMING FACTOR**, A FACTOR NOW EXPLAINED AS THE EXODUS OF DNA FROM CELL TO CELL.

WILLIAM ASTBURY

IN 1937 PHYSICIST-BIOLOGIST **WILLIAM ASTBURY** CREATED THE FIRST X-RAY DIFFRACTION PATTERNS, THUS DEMONSTRATING THAT DNA POSSESSES A REGULAR STRUCTURE.

AVERY, ET AL

ASTBURY'S WORK PAVED THE WAY FOR 1943 EXPERIMENTS BY **OSWALD AVERY**, **COLIN MACLEOD** AND **MACLYN MCCARTY**, WHICH SINGLED OUT DNA AS THE TRANSFORMING FACTOR IN GRIFFITH'S PNEUMOCOCCUS EXPERIMENTS.

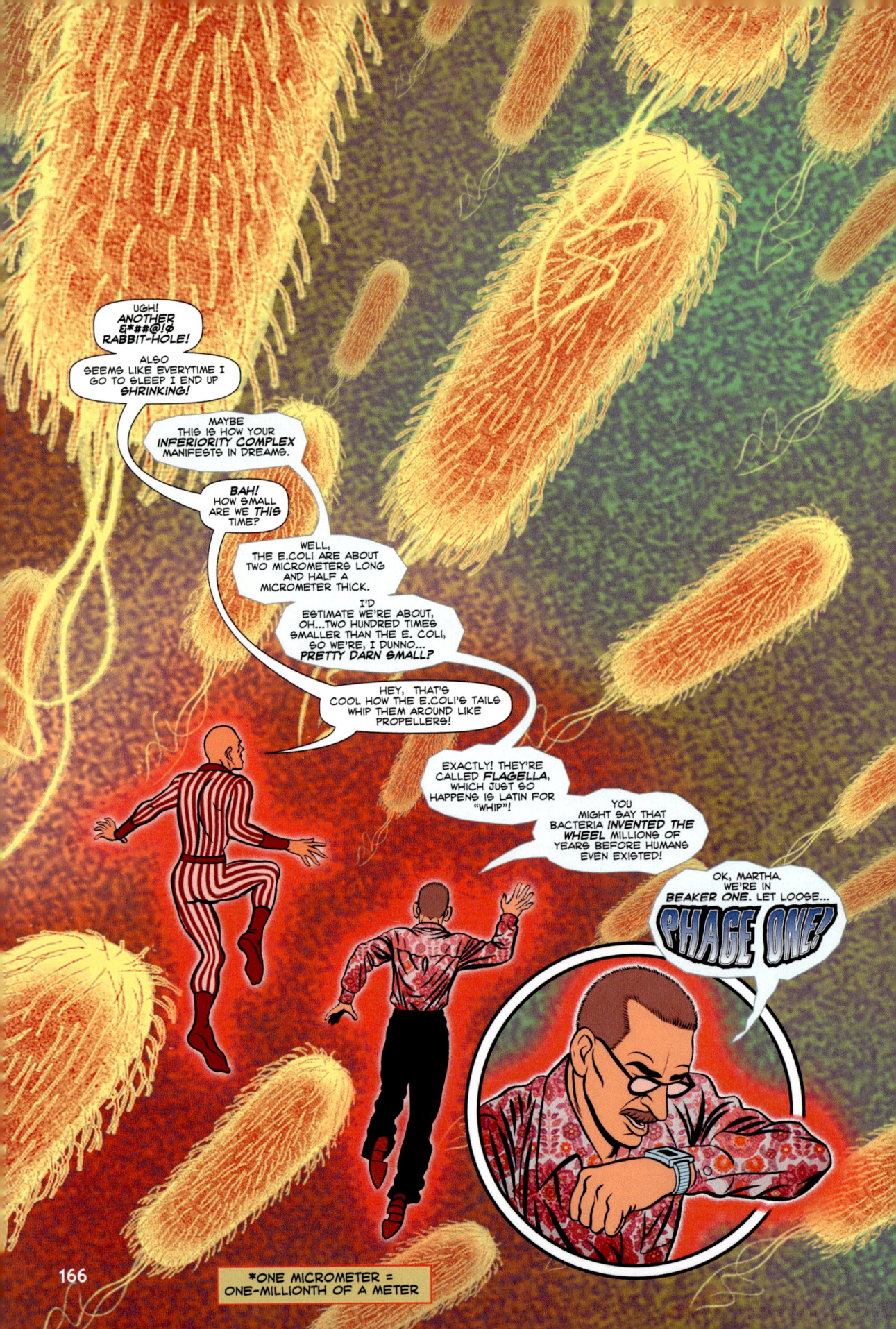

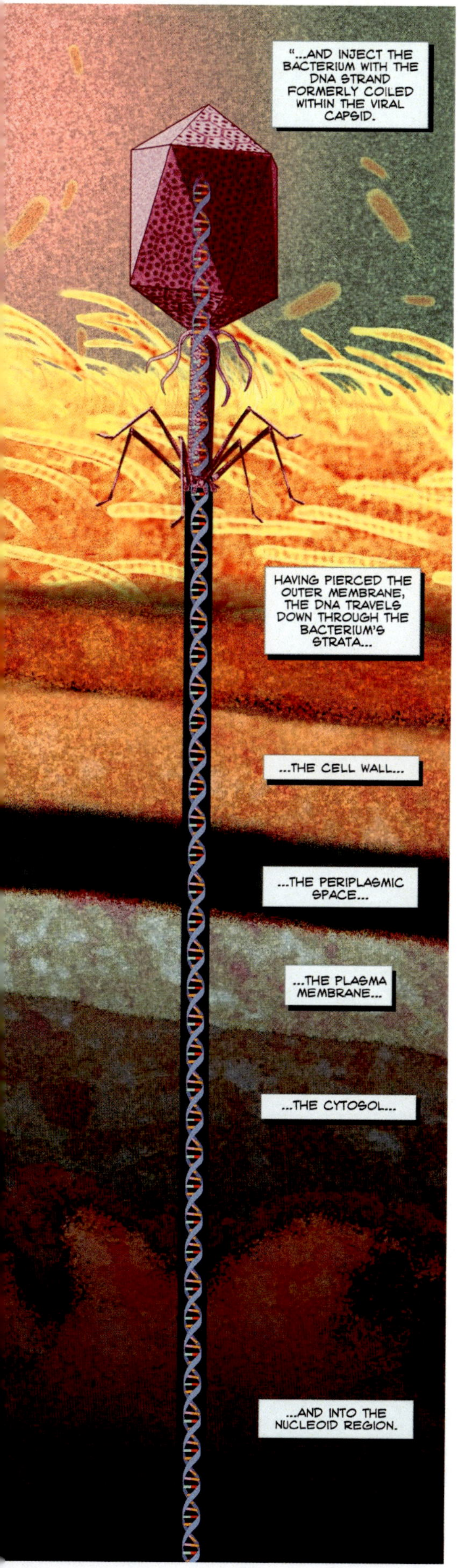

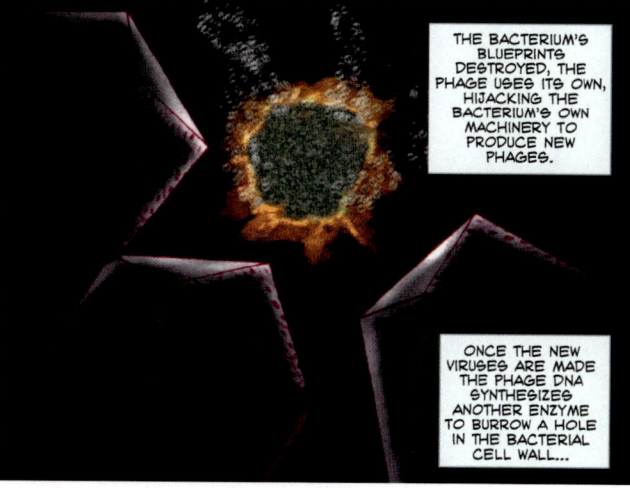

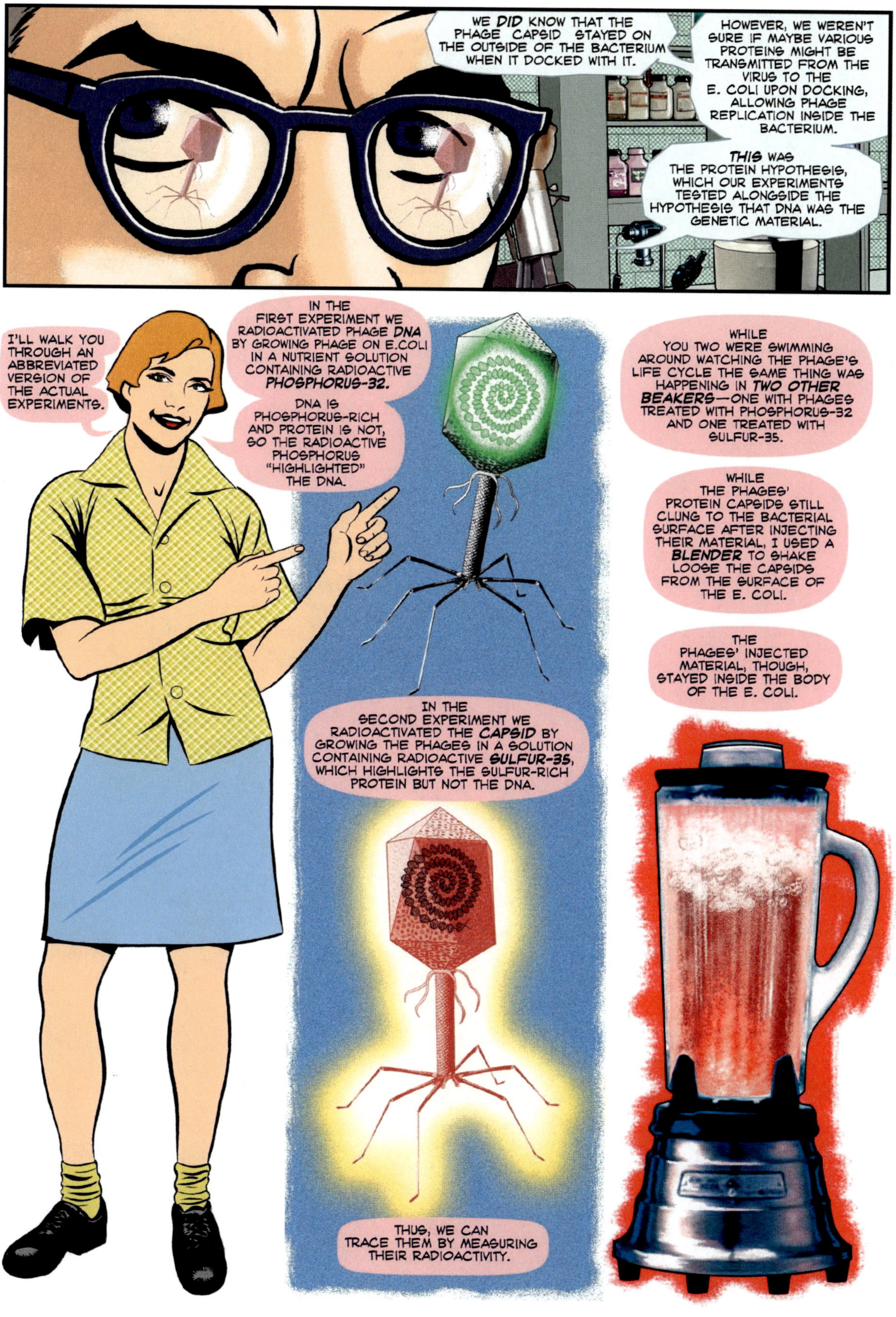

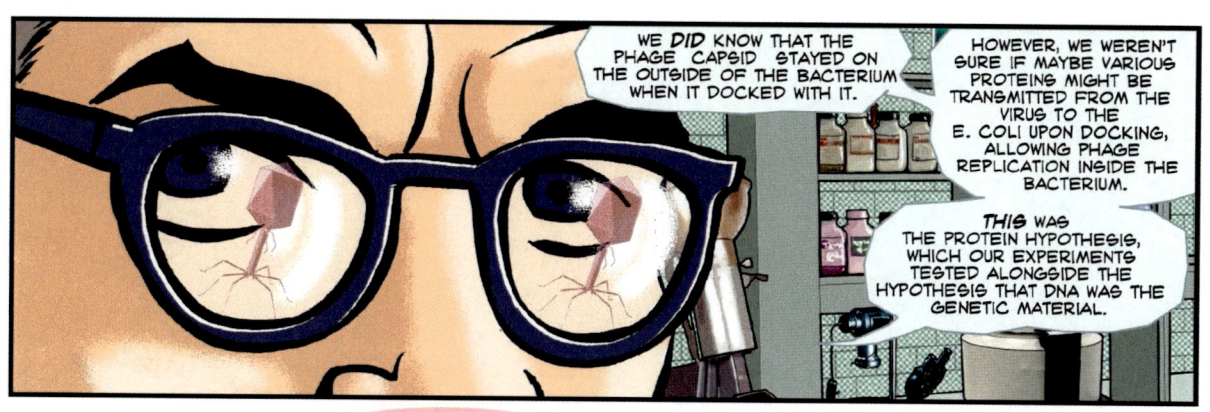

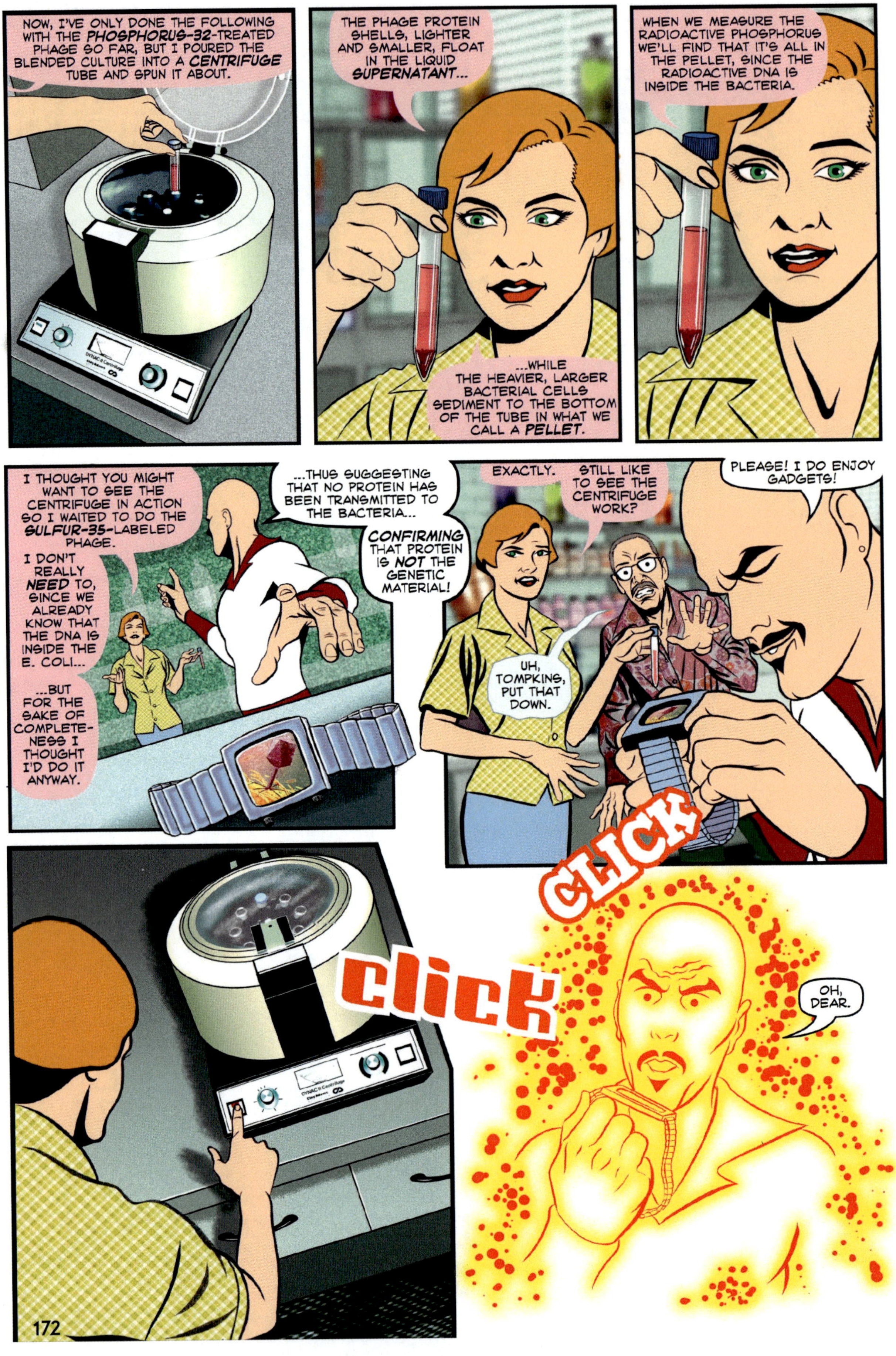

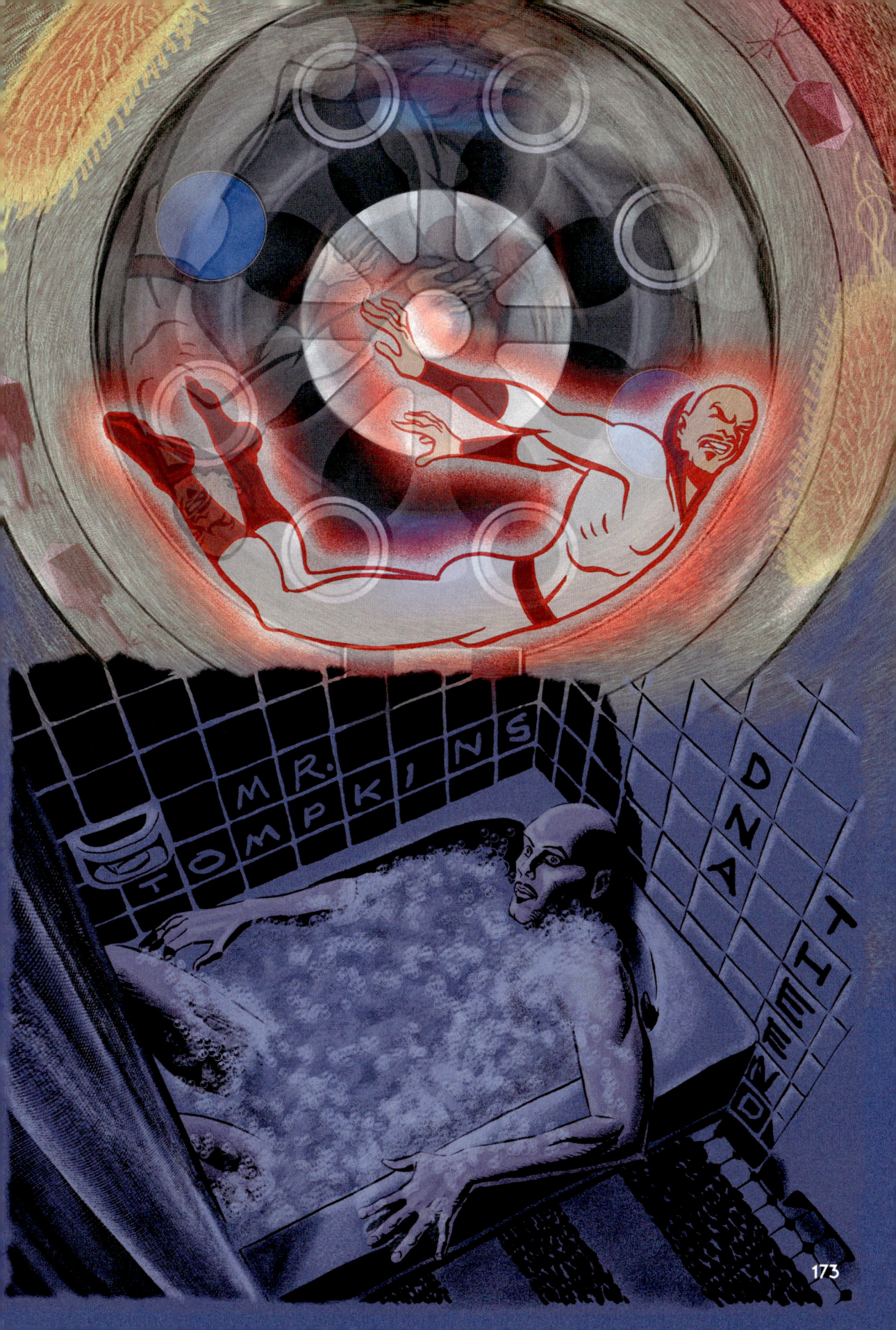

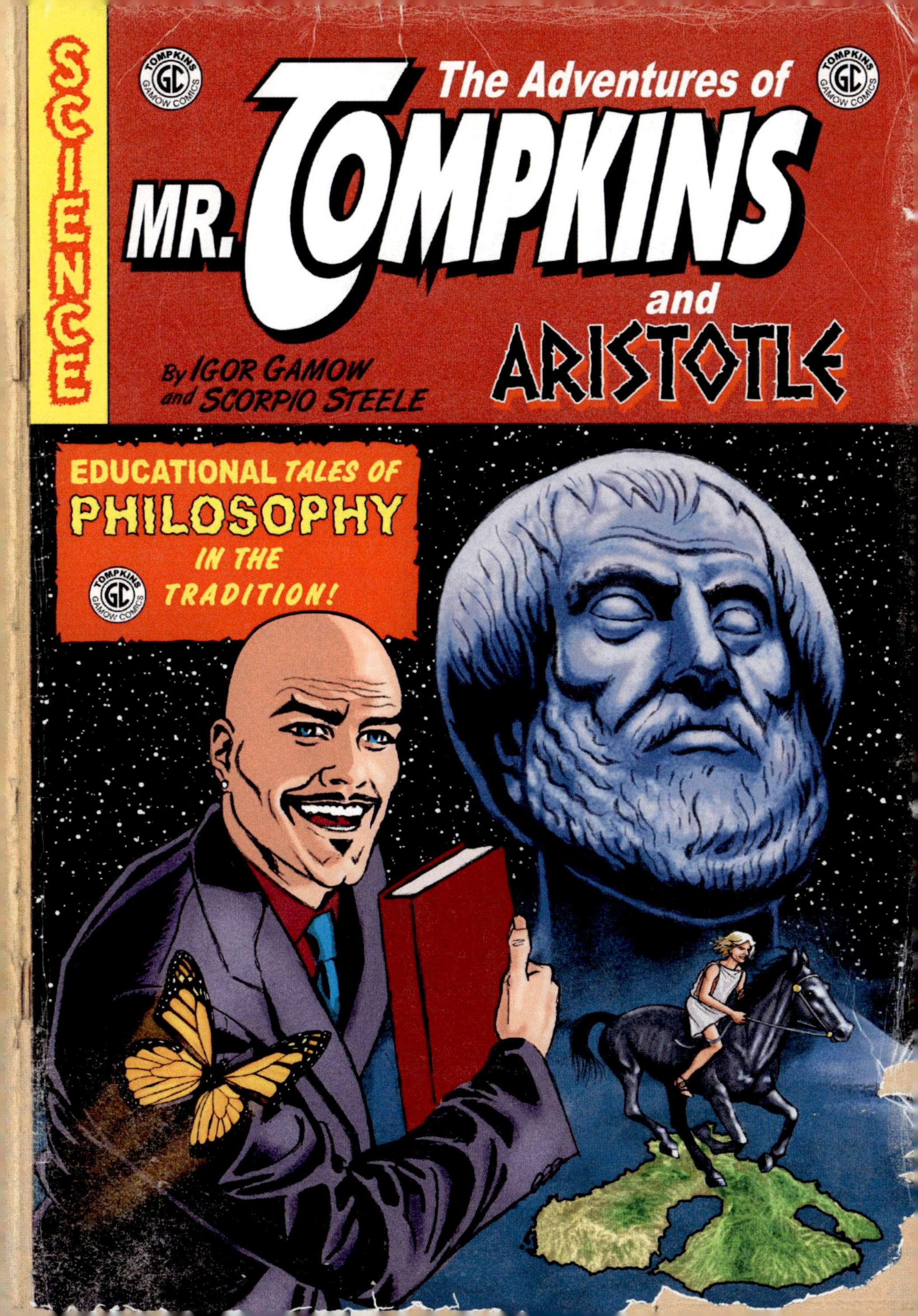

"Some persons move in their sleep, and perform many acts like waking acts, but not without a phantasm or an exercise of sense-perception; for a dream is in a certain way a sense-impression."

Aristotle
On Sleep and Sleeplessness

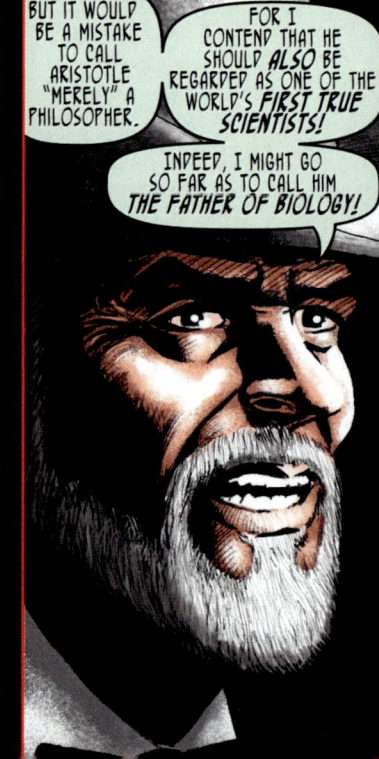

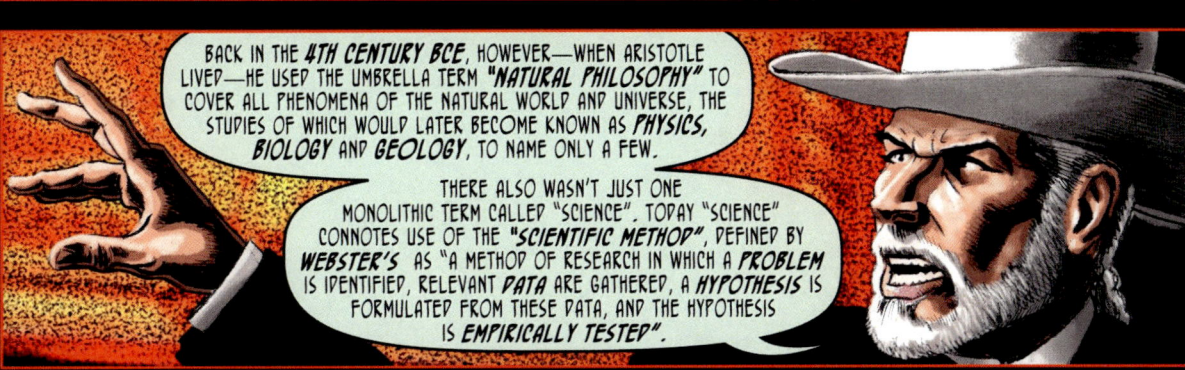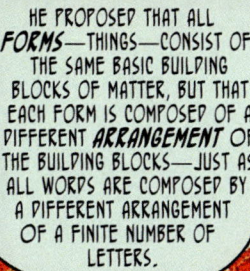

To walk with me you must know certain principles and causes—how things work. You must know something about everything, though you need not know it in great detail.

I also require you to learn difficult-to-learn information. Everyone has sense-perception, so anything gleaned merely through the senses is easy and thus, not wisdom.

The person with the highest degree of universal knowledge of things difficult to learn, and thus farthest from a simple sense-perception, is the wisest.

Lastly, you can't call yourself theoretically wise unless you know your subject well enough to teach it to someone else.

However, theoretical knowledge is not enough—you need practical wisdom, too, so you can deliberate well about what sorts of things are going to enable you to navigate your way pleasantly through life.

While it is important to have knowledge for its own sake, if you do not also have practical wisdom to complement this scientific knowledge, then happiness may elude you.

For example, you may be a Stock Exchange Sorcerer, amassing all the riches of the world, but a lack of practical knowledge on how best to comport yourself might make you many enemies on the way up the money ladder. Misery would be your only companion.

YOU KNOW, TOMPKINS, YOUR MENTION OF THE **TAO TE CHING** REMINDS ME OF ANOTHER SOURCE OF PRACTICAL WISDOM--BENJAMIN HOFF'S **THE TAO OF POOH**, WHEREIN HOFF USES **WINNIE THE POOH** CHARACTERS TO EXPLAIN TAOISM.

"**POOH**" PUNS ON THE CHINESE **P'U**--THE "**UNCARVED BLOCK**"--THE ORIGINAL SIMPLICITY IN WHICH SOMETHING OR SOMEONE EXISTS BEFORE COMPLEXITY AND ARROGANCE COME ALONG AND RUIN IT.

THE PERSON WHO VALUES "KNOWLEDGE FOR KNOWLEDGE'S SAKE" HOFF CALLS THE "**CONFUSIONIST, DESSICATED SCHOLAR**" AND LABELS **OWL** AS SUCH.

SUCH SCHOLARS, HE SAYS, SHOULD "GO OUTSIDE AND SNIFF AROUND--WALK THROUGH THE GRASS, **TALK TO THE ANIMALS**", RATHER THAN LEARN A MESS OF FACTS ONLY FROM BOOKS.

I MAY **SOUND** AT TIMES LIKE A CONFUSIONIST, BUT I **CAN SAY** THAT I **DO** GO OUTSIDE WITH THE ANIMALS...

...THOUGH I DON'T SO MUCH AS **TALK** TO ANIMALS AS, WELL... **VIVISECT** THEM.

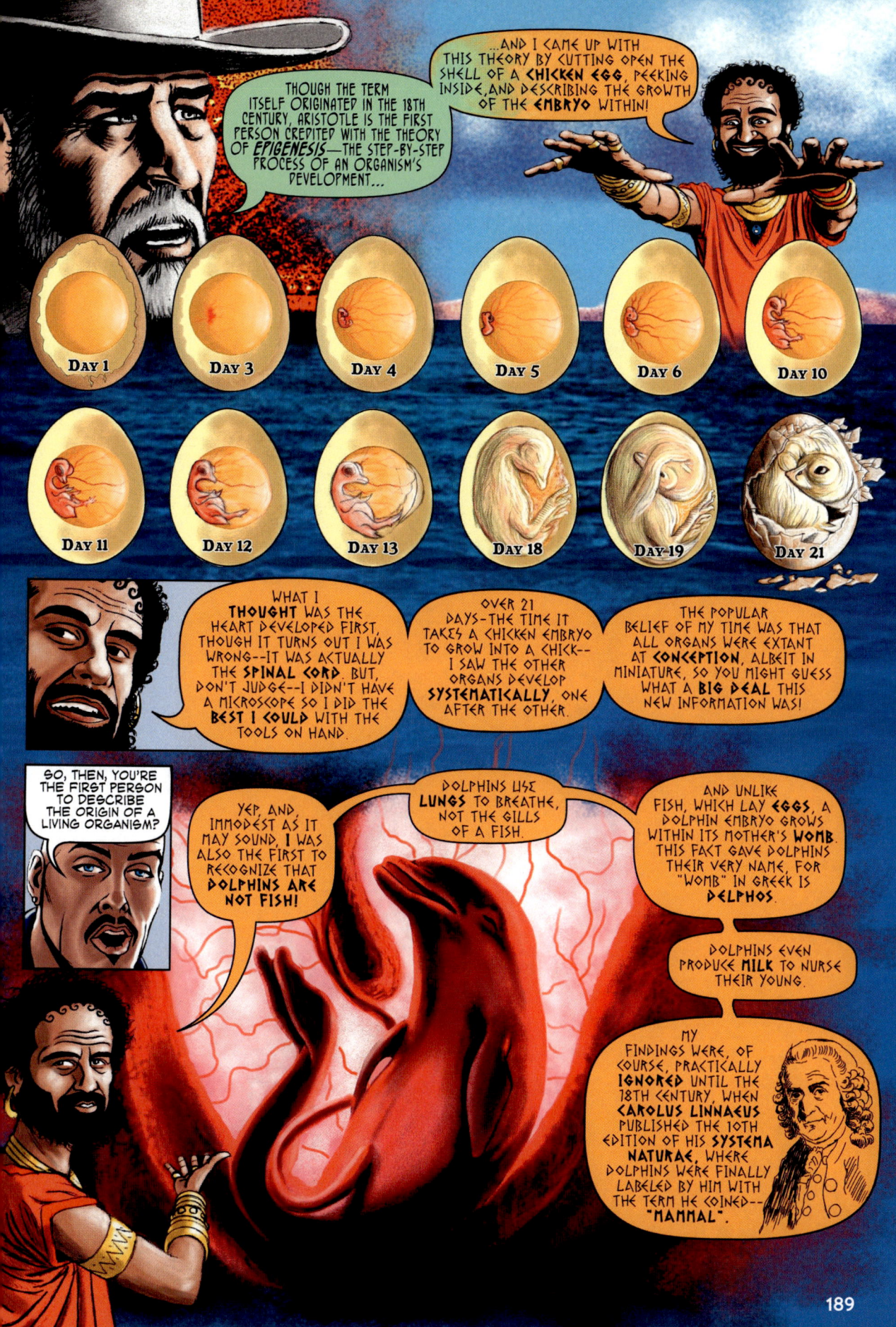

"ORPHANED, I WENT TO LIVE FOR SEVEN YEARS WITH MY SISTER AND HER HUSBAND IN ATARNEUS, A CITY ON THE COAST OF ASIA MINOR, JUST EAST OF LESBOS.

"DURING THIS TIME I STUDIED DILIGENTLY, AND SINCE I WAS A PRECOCIOUS STUDENT--**AND** BECAUSE MY FATHER HAD BEEN PART OF THE ROYAL COURT--I WAS ADMITTED TO **THE ACADEMY** IN ATHENS!

"I WAS DISAPPOINTED WHEN I GOT THERE, THOUGH, BECAUSE HE FROM WHOM I MOST WANTED TO LEARN WAS AWAY IN **SYRACUSE**, TUTORING ITS RULER.

"BUT THEN, TWO YEARS INTO MY STUDIES AT THE ACADEMY, I FINALLY GOT TO LEARN WITH THE MASTER HIMSELF... **PLATO**!

"ONCE HE'D BEEN A WRESTLER AND HAD TAKEN THE MONIKER "PLATO" TO DISTINGUISH HIMSELF. BORN **ARISTOCLES**, HIS NEW NAME DERIVED FROM THE GREEK WORD FOR "BROAD", AND SUITED HIM BETTER. FOR GREAT WERE THE BREADTHS OF BOTH HIS SHOULDERS AND OF HIS KNOWLEDGE."

THE DIRECTION IN WHICH EDUCATION STARTS A MAN WILL DETERMINE HIS FUTURE LIFE.

"SINCE PLATO HAD SUCH A HUGE IMPACT ON MY LIFE, I SUPPOSE I WOULD BREAK MY BIOGRAPHY INTO THREE PARTS--'**PRE-PLATO**', WHICH WE JUST SAW, '**PLATO**', WHICH WE NOW ENTER, AND THE STILL-TO-COME '**POST-PLATO**'.

"ANYHOW, ALTHOUGH I **RESPECTED** PLATO AS MY MENTOR, OUR RELATIONSHIP OVER THE YEARS WAS OFTEN **TUMULTUOUS**. YOUNG AND REBELLIOUS, MY OPINIONS EVENTUALLY **DIVERGED** FROM HIS."

I TELL YOU, A PERSON'S FUNCTION IS DELIBERATION, RULING, LIVING, AND TAKING CARE OF THINGS!

NO. A PERSON SHOULD PERFORM ACTIVITIES WHICH EXPRESS **REASON**.

THE IDEAL SOCIETY IS AN UTOPIA RUN BY AN ELITE OF **PHILOSOPHER KINGS** WHO SHARE FAMILY AND PROPERTY-- OWNERSHIP AND PROFIT SHOULD BE **ELIMINATED** FROM LIFE

THIS ELITE WOULD OVERSEE THE LOWER CLASSES-- SOLDIERS AND COMMONERS.

SOCIETY WORKS BEST IF EVERYONE **KNOWS** THEIR PLACE IN IT.

NO. OWNING SOMETHING **YOURSELF** MAKES AN ENORMOUS DIFFERENCE IN THE PLEASURE IT PROVIDES, AND ENCOURAGES PEOPLE TO DO BETTER DEEDS.

OH, GO HOME TO YOUR READING SHOP, YOU **MIND ON LEGS**! AH-HAHAHAHA

SORRY, PLATO, YOU'RE DEAR TO ME, BUT DEARER STILL IS THE **TRUTH**! HA, HA!

"PLATO AND I EVEN BUTTED HEADS ON THE NATURE OF REALITY ITSELF. HE THOUGHT AN ULTIMATE REALITY EXISTS **TRANSCENDENT** FROM THE MATERIAL WORLD, COMPOSED OF PERFECT AND ETERNAL VERSIONS OF ALL OBJECTS EXTANT IN OUR ILLUSORY MATERIAL WORLD. FOR EXAMPLE, IN THIS ULTIMATE REALITY WOULD EXIST **THE PERFECT BUTTERFLY**, FLAWLESS AND IMMORTAL.

"WHEREAS, IN OUR EVERYDAY, ILLUSORY **MATERIAL** WORLD WE GET ONLY A LOW-RENT **COPY** OF THAT PERFECT BUTTERFLY, ONE WHICH WILL LIVE OUT ITS SHORT LIFE AND DIE. NO MATTER HOW HEALTHY OR BEAUTIFUL A BUTTERFLY FROM OUR MATERIAL WORLD MAY BE, IT CAN NEVER BE AS GOOD AS THE **IDEA** OF THE PERFECT BUTTERFLY ITSELF.

"I, HOWEVER, ARGUED THAT **THINGS**--OBJECTS MADE OF MATTER--ARE WHAT'S REAL, AND IF ONE IS TO UNDERSTAND A THING, THEN ONE NEEDS TO KNOW ITS FOUR **"CAUSES"**: MATERIAL, FORMAL, EFFICIENT, AND FINAL.

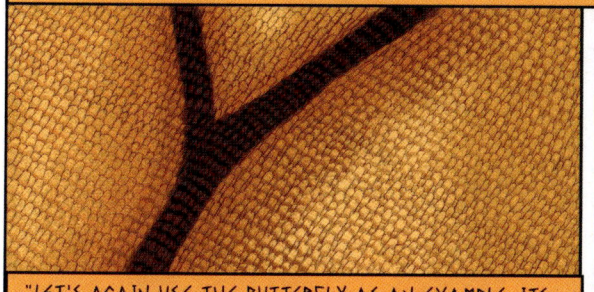

"LET'S AGAIN USE THE BUTTERFLY AS AN EXAMPLE. ITS **MATERIAL** CAUSE IS THE **MATTER** OF WHICH IT'S MADE: CELLS, SCALES, PIGMENTS AND CHEMICALS.

"ITS **FORMAL** CAUSE IS THE ESSENCE OR 'BLUEPRINT' INTO WHICH IT'S MADE, ITS **'BUTTERFLY-NESS'**.

"**EFFICIENT** CAUSE: THE SOURCE BY WHICH IT'S MADE--ITS **LIFE-CYCLE**...

"...WHICH BEGINS WITH AN EGG, PROTECTED BY A THIN LAYER OF WAX SO THAT IT DOESN'T DRY OUT BEFORE...

"...HATCHING INTO A **CATERPILLAR**, CHEWING LEAVES FOR SUSTENANCE...

"...BEFORE FINDING THE UNDERSIDE OF A LEAF WHERE IT CAN TRANSFORM ITSELF, FIRST INTO A **CHRYSALIS**..."

...AND FINALLY, INTO A **BUTTERFLY!**

IN GENERAL, **21ST CENTURY** SCIENCE IS MOST CONCERNED WITH EFFICIENT CAUSE.

LASTLY, **FINAL** CAUSE IS THE **PURPOSE** FOR WHICH A THING'S MADE.

SEEMS LIKE FINAL CAUSE WOULD BE EASIER TO DETERMINE FOR NON-LIVING THINGS, SUCH AS A WHEEL. BUT A BUTTERFLY'S PURPOSE?

THAT'S OPEN TO INTERPRETATION, RIGHT?

"So I guess this brings us to the **'post-Plato'** period of my life.

"After twenty years at the Academy I couldn't help but hope that I would succeed Plato as Academy director.

"But it wasn't to be-- the divide between our philosophies was too great, and before he died in **347 BCE** Plato appointed his nephew, **Speusippus**. So I packed my bags and prepared for a move...

CHANGE IN ALL THINGS IS SWEET.

FAREWELL, ATHENS.

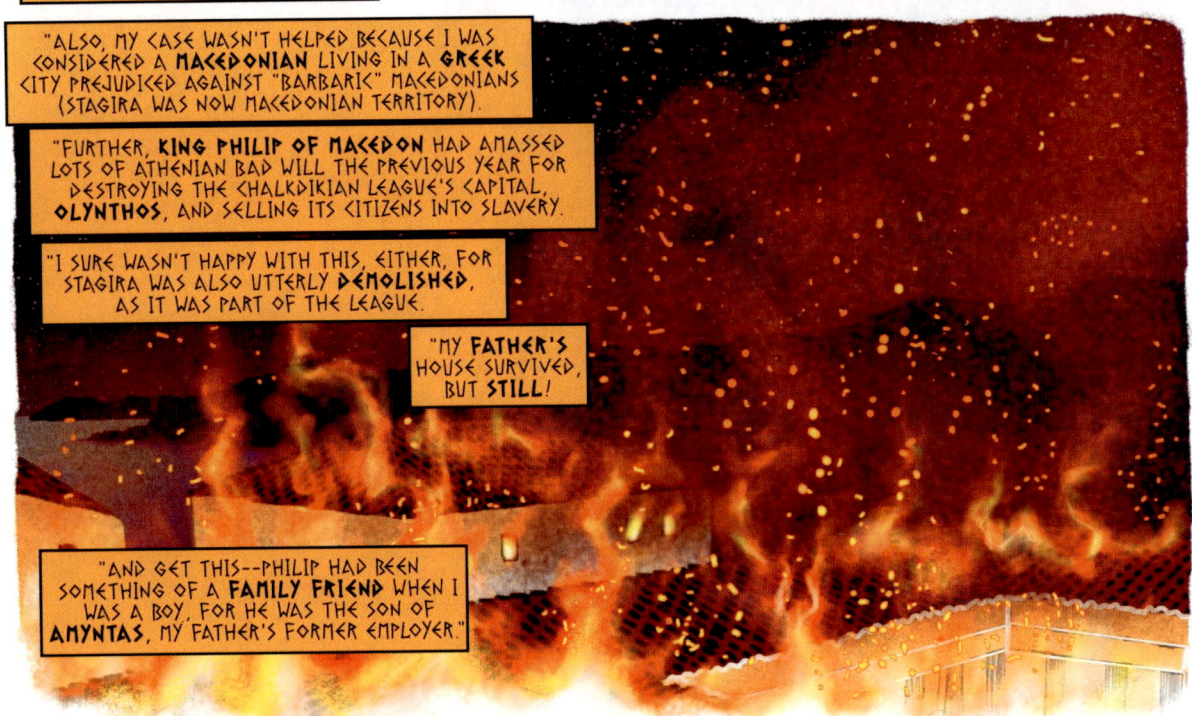

"Also, my case wasn't helped because I was considered a **Macedonian** living in a **Greek** city prejudiced against "barbaric" Macedonians (Stagira was now Macedonian territory).

"Further, **King Philip of Macedon** had amassed lots of Athenian bad will the previous year for destroying the Chalkidian League's capital, **Olynthos**, and selling its citizens into slavery.

"I sure wasn't happy with this, either, for Stagira was also utterly **demolished**, as it was part of the League.

"My **father's** house survived, but **still!**

"And get this--Philip had been something of a **family friend** when I was a boy, for he was the son of **Amyntas**, my father's former employer."

YOUR OLD FAMILY FRIEND DESTROYED YOUR HOMETOWN?

INDEED. I GUESS THAT IT WAS "JUST BUSINESS", AND "NOTHING PERSONAL," YOU KNOW? WHATEVER.

ANYWAY, HERE WE ARE IN **ATARNEUS**, WHERE I'D LIVED PRIOR TO ATTENDING THE ACADEMY.

AT THE BEHEST OF PHILIP, THE PHILOSOPHER **XENOCRATES** AND I SAILED HERE TO VISIT THE ROYAL COURT OF ANOTHER ACQUAINTANCE, **HERMIAS OF ATARNEUS**.

ATARNEUS WAS ON THE RISE AND PHILIP WANTED ME TO HELP ESTABLISH AN **ALLIANCE** BETWEEN IT AND MACEDON.

ATARNEUS' LOCATION WAS TACTICALLY ADVANTAGEOUS FOR CAMPAIGNS AGAINST THRACE AND PERSIA.

I LIKED HERMIAS. HE'D STUDIED AT THE ACADEMY, TOO, AND SEEMED TO WANT TO BE THE SORT OF **'PHILOSOPHER KING'** ABOUT WHICH PLATO HAD SPOKEN.

"AT NEARBY **ASSOS**, HERMIAS SET ME UP WITH MY OWN ACADEMY... AND A **WIFE**! HIS NIECE AND ADOPTED DAUGHTER, **PYTHIAS**.

"TWO YEARS LATER, IN **345 BCE**, WE MOVED A SHORT JAUNT SOUTH TO **LESBOS** AND SETTLED IN TO FAMILY LIFE AT **MYTILENE**.

"THE TWO YEARS IN LESBOS WERE PROBABLY THE BEST YEARS OF MY LIFE! I HAD A GORGEOUS WIFE AND DAUGHTER, AND MY BOTANIST PUPIL, **THEOPHRASTOS**, AND I HAD ENOUGH FORMS OF MARINE LIFE TO KEEP A MAN DEEP IN STUDY FOR THE REST OF HIS LIFE.

"BUT, ALL GOOD THINGS MUST END, AND BEFORE LONG I ACCEPTED KING PHILIP'S INVITATION TO GO TO **PELLA**, THE MACEDONIAN CAPITAL, TO TUTOR HIS 13-YEAR-OLD SON, **ALEXANDER**. I DIDN'T REALLY WANT TO, BUT I ACCEPTED UNDER THE CAVEAT THAT PHILIP REBUILD STAGIRA.

"SO, AT THE **TEMPLE OF THE NYMPHS** IN THE NEARBY VILLAGE OF **MIEZA**, I INSTRUCTED ALEXANDER AND HIS FRIENDS, WHICH INCLUDED **CASSANDER** AND **PTOLEMY**, BOTH OF WHOM EVENTUALLY BECAME MONARCHS IN THEIR OWN RIGHT.

WAIT—THAT'S **ALEXANDER THE GREAT**, RIGHT? THE WORLD CONQUEROR!

YES, I TUTORED HIM OVER A PERIOD OF...OH, MAYBE THREE YEARS?

"IN ADDITION TO PHILOSOPHY I TAUGHT ALEXANDER WHAT LITTLE MY FATHER HAD IMPARTED TO ME OF **MEDICINE**.

"ALEXANDER LOVED HAVING THE ABILITY TO PRESCRIBE TREATMENTS FOR SICK FRIENDS. ALTHOUGH HE COULD BE BAD-TEMPERED, HE WAS GRACIOUS TO THOSE WHOM HE CONSIDERED COMRADES.

"HE **LOVED** TO READ. HIS FAVORITE STORY WAS HOMER'S **THE ILIAD**, WHICH HE REGARDED AS A **MANUAL** ON THE ART OF WAR. I PREPARED AN ANNOTATED VERSION FOR HIM, AND HE ALWAYS KEPT ONE OF THE SCROLLS FROM IT, ALONG WITH A DAGGER, UNDER HIS PILLOW.*

"LATER, ON CAMPAIGNS, HE EVEN KEPT **THE ILIAD** IN AN ARMORED BOX TO PROTECT IT FROM HARM.

*IN ARISTOTLE'S DAY A "BOOK" WOULD HAVE BEEN A COLLECTION OF NUMEROUS SCROLLS MADE OF PARCHMENT. --EDITOR

"SPEAKING OF BONDS, ALEXANDER ONCE SAID THAT HE LOVED ME LIKE A FATHER...

ARISTOTLE, PHILIP MAY HAVE GIVEN ME LIFE... BUT YOU HELP ME TO **LIVE WELL** BY INSTILLING IN ME A THIRST FOR KNOWLEDGE.

"...YET YEARS LATER WE EVENTUALLY GREW DISTANT AND HAD A **FALLING OUT**. SINCE I'D RATHER KEEP THINGS POSITIVE I'D RATHER NOT GO INTO THAT, Y'KNOW? IF YOU REALLY NEED TO KNOW YOU CAN READ **PLUTARCH'S** "LIFE OF ALEXANDER", WHERE IT'S PRETTY WELL EXPLAINED.

"SO...FROM PELLA I MOVED BACK TO STAGIRA FOR A LITTLE WHILE AND THEN AGAIN BACK TO ATHENS, WHERE I SPENT THE NEXT THIRTEEN YEARS.

"I WAS **FIFTY** WHEN, BESIDE A GROVE DEDICATED TO **APOLLO LYCEUS**-- "APOLLO THE WOLF GOD"-- I FOUNDED MY OWN VERSION OF THE ACADEMY-- **THE LYCEUM**.

"THE LYCEUM BUILDING ITSELF HAD BEEN AROUND FOR YEARS BEFORE THIS, AND WAS A SITE OF PHILOSOPHICAL DISCUSSION, MILITARY TRAINING AND ATHLETICS. BUT I MADE IT OFFICIAL.

"IN THE MORNING WE'D WALK THROUGH THE GROVE AND I'D LECTURE ON ALL THE USUAL SUBJECTS-METAPHYSICS, BIOLOGY, ETHICS, POLITICS, ART, POETICS...YOU NAME IT.

"ONCE ALEXANDER SUCCEEDED PHILIP AS KING OF MACEDON AND WAS OFF CONQUERING ASIA, HE'D SHIP US ANIMAL AND PLANT SPECIMENS TO STUDY. WITH THESE SPECIMENS AND THE MONEY HE SENT WITH THEM I WAS EVEN ABLE TO OPEN THE FIRST-EVER **ZOO** AND **BOTANICAL GARDENS**.

"DURING THIS PERIOD OF TIME MY WIFE PASSED AWAY. SHE MEANT SO MUCH TO ME THAT A PROVISION OF MY WILL WAS THAT HER BONES BE **DISINTERRED** AND **REBURIED** WITH MY OWN WHEN I DIED.
(...SIGH...)
SHE DIED SO **YOUNG**...

"THEN, IN **323 BCE** MY TIME AT THE LYCEUM DREW TO A CLOSE AFTER ALEXANDER DIED OF FEVER IN BABYLON.

"ANTI-MACEDONIAN SENTIMENT SWEPT ATHENS. MY CONNECTION WITH THE 'TYRANT' ALEXANDER WAS WELL-KNOWN, SO I WAS **GUILTY BY ASSOCIATION**.

"I ENDED UP ACCUSED OF **'IMPIETY'** FOR A SONG I HAD ONCE COMPOSED IN HONOR OF HERMIAS WHEN HE DIED. IN IT, I SUPPOSEDLY COMPARED HERMIAS TO A GOD... **WHATEVER**.

♪...HIS DEATHLESS NAME, HIS PURE CAREER, LIVE SHRINED IN SONG, AND LINK'D WITH AWE, THE AWE OF XENIAN JOVE, AND FAITHFUL FRIENDSHIP'S LAW.♪

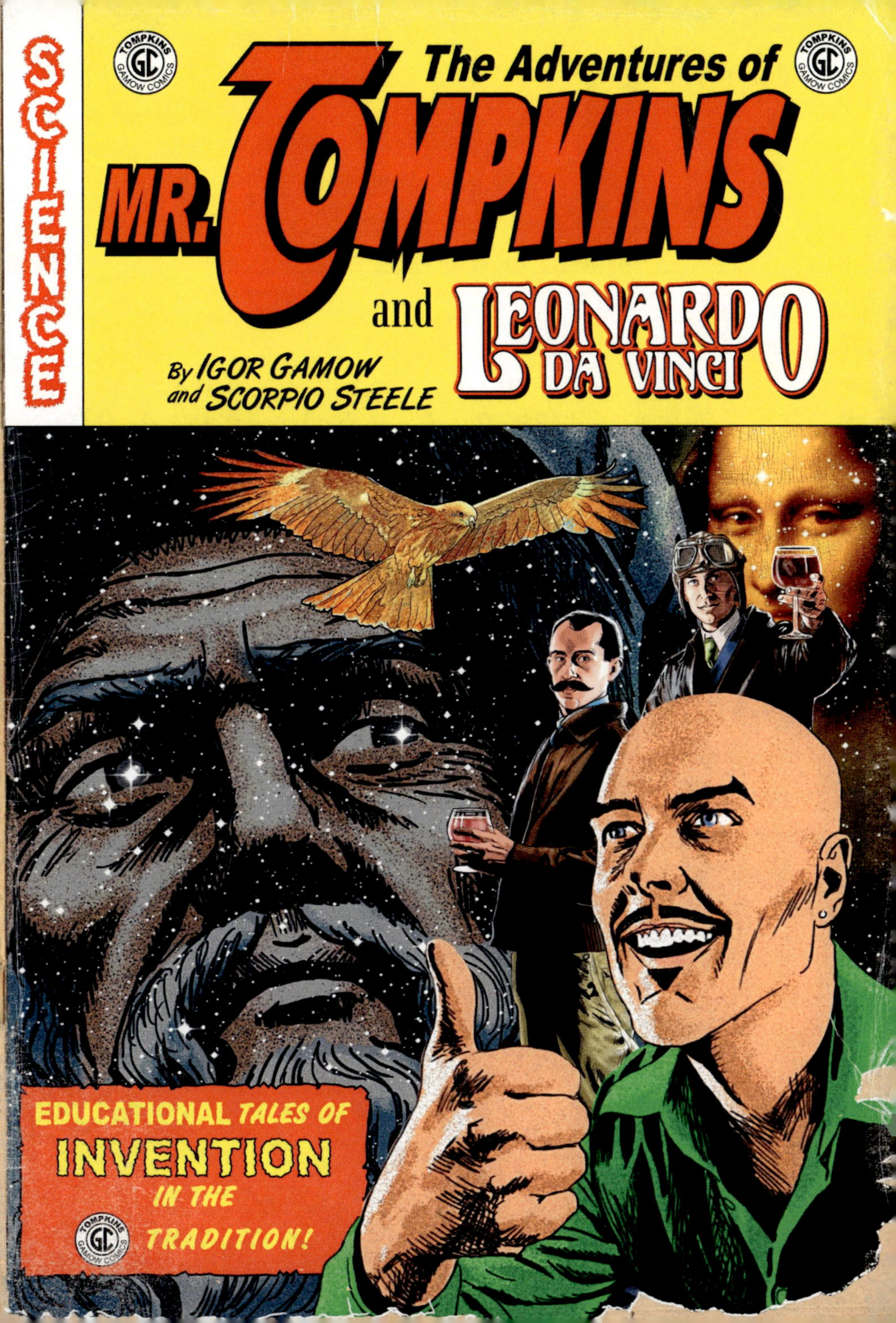

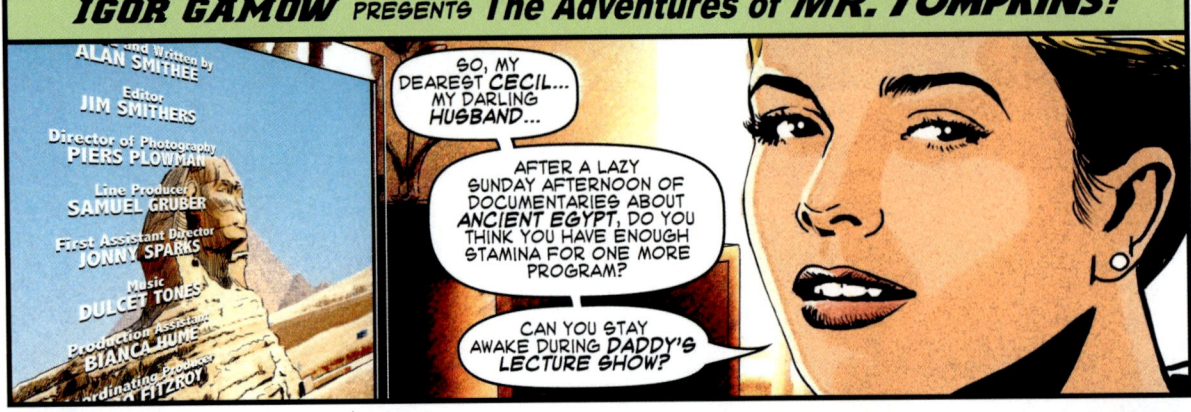

CORRECT! *NO ONE* is buried in Grant's tomb. Burial is *BELOW* ground and a tomb is *ABOVE* ground.

Since you're so smart, you'll know the answer to the second question—who painted the *MONA LISA?*

LEONARDO DA VINCI.

LEONARDO DA VINCI.

That's right! *LEONARDO DA VINCI!* However, I bet that you *DON'T* know why the *MONA LISA*, a painting by an Italian, is displayed in the *LOUVRE*, a *FRENCH* museum!

Well, we'll get to that in a moment, but I *ALSO* ask you why, among all the other famous Renaissance painters, is Leonardo the *MOST FAMOUS* of them all?

The answer to that can be found in the handful of legendary paintings, beginning, of course, with the *MONA LISA*...

"...BUT THERE IS STILL, OF COURSE, THE LAST SUPPER..."

"...THE VIRGIN AND CHILD WITH ST. ANNE..."

"...AND LADY WITH AN ERMINE", AMONG OTHERS."

"BUT LABELING LEONARDO AS JUST A PAINTER WOULD DO HIM A DISSERVICE, FOR PAINTING WAS A RELATIVELY SMALL PART OF HIS TOTAL OUTPUT.

LEONARDO'S MODERN STATUS AS A GENIUS CAN BE ATTRIBUTED TO THE WRITINGS AND SKETCHES FROM HIS NOTEBOOKS.

ONLY 7,000 OF 13,000 TOTAL PAGES REMAIN, BUT FROM THESE SURVIVING DOCUMENTS IT IS PLAIN THAT LEONARDO WAS, IN ADDITION TO BEING A WONDERFUL PAINTER, ALSO A MASTER ARCHITECT, ENGINEER, AND INVENTOR."

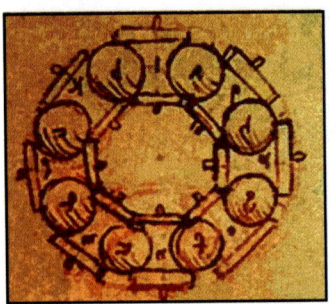

"AMONG HIS INVENTIONS PROVEN TO WORK BY MODERN SCIENCE ARE THE BALL BEARING AND THE PARACHUTE..."

"HE PERFORMED NUMEROUS EXPERIMENTS INTO THE NATURE OF OPTICS AND PROPOSED, FOR EXAMPLE, THAT LIGHT IS A WAVE 200 YEARS BEFORE THE IDEA'S FORMULATION BY CHRISTIAN HUYGENS..."

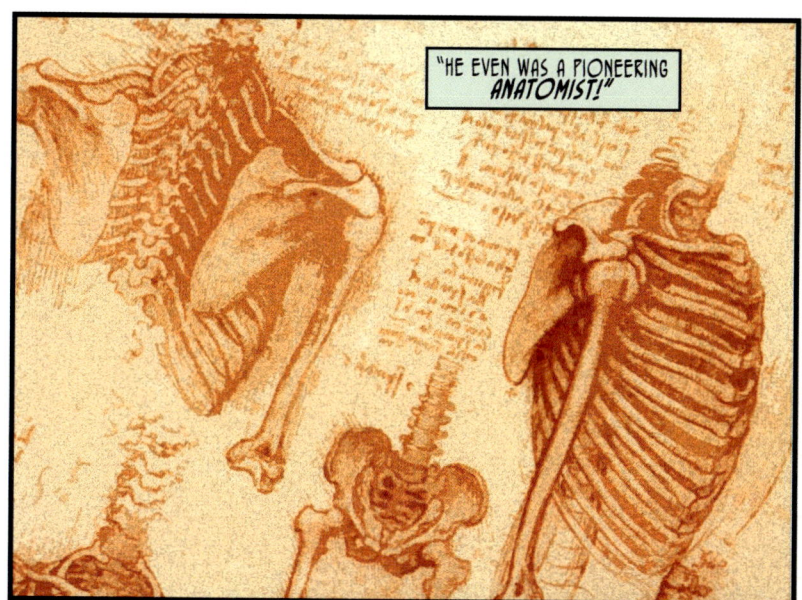

"Why does the eye see a thing more clearly in dreams than the imagination when awake?"

Leonardo da Vinci

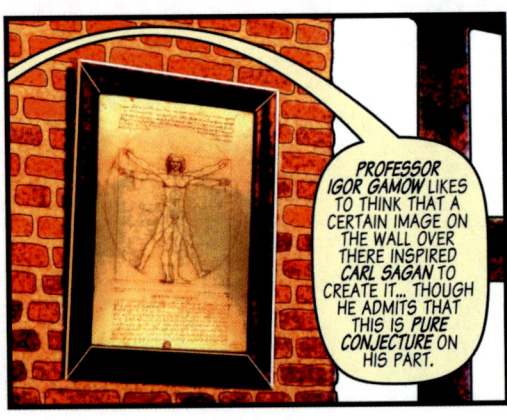

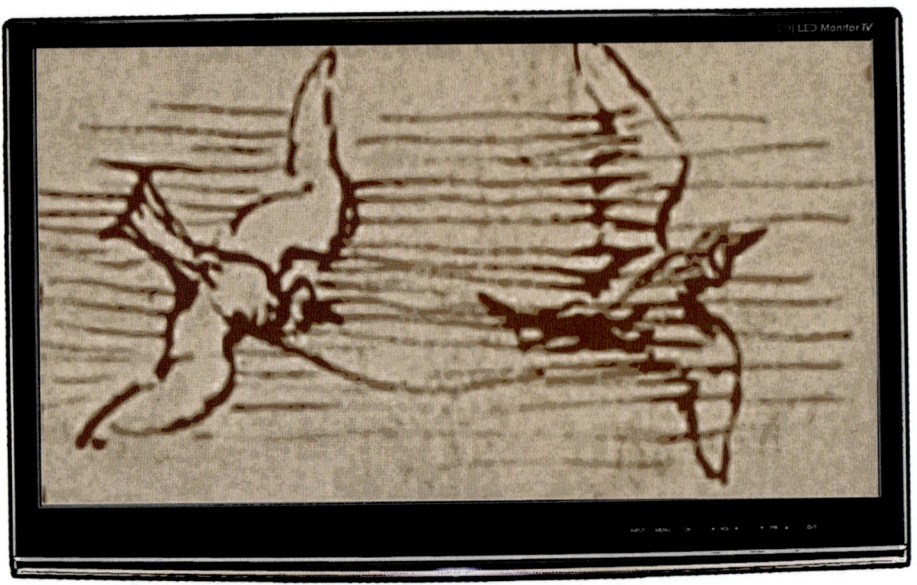

"LEONARDO, YOU WERE THE FIRST PERSON TO DRAW *HORIZONTAL LINES* TO REPRESENT THE *WIND* AND ITS EFFECTS ON A BIRD IN FLIGHT...

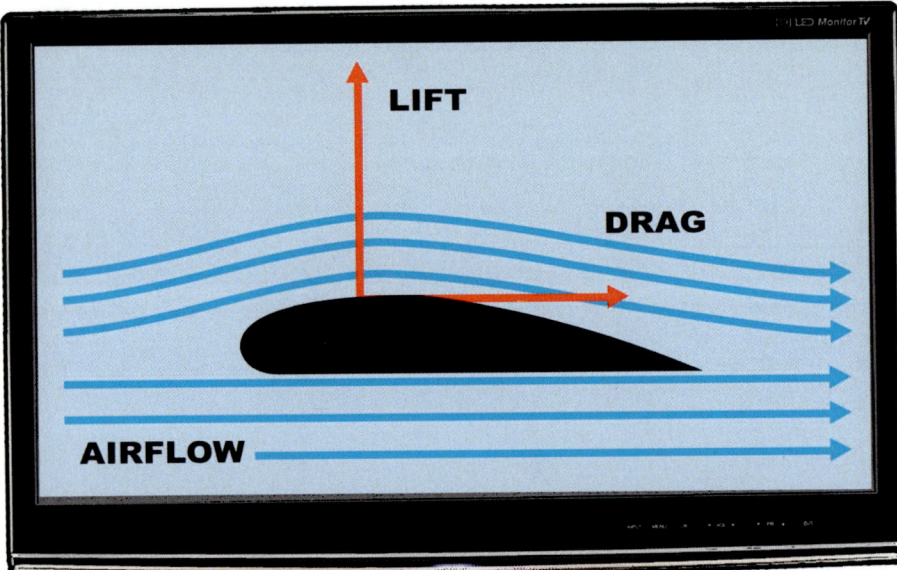

"...A TECHNIQUE STILL USED IN THE MODERN ERA, AS YOU CAN SEE IN THIS DIAGRAM OF A WING CUTTING THROUGH THE AIR.

"AIR VELOCITY *INCREASES* WHEN IT FLOWS UNEVENLY OVER THE CURVED *UPPER* SURFACE OF A WING, REDUCING PRESSURE ABOVE THE WING AND CREATING *LIFT*.

"THIS IS CALLED *BERNOULLI'S PRINCIPLE*, AFTER THE SWISS MATHEMATICIAN WHO FIRST DESCRIBED IT, *DANIEL BERNOULLI*.

"AND NOW HERE WE HAVE A COUPLE OF THE FLYING MACHINES THAT OUR ILLUSTRIOUS *HOST* DESIGNED.

LEONARDO, OBVIOUSLY YOU RECOGNIZE THESE-- YOUR 1490 *GLIDER* AND YOUR 1493 *"HELICAL AIR SCREW"*.

TOMPKINS, THE IDEA BEHIND THE AIR SCREW WAS TO TO COMPRESS AIR TO ACHIEVE FLIGHT. THE FIRST DRAWING OF A "HELICOPTER", YOU COULD SAY.

 "OKAY, SO NOW WE JUMP FORWARD A FEW HUNDRED YEARS, AND HERE WE HAVE *SIR GEORGE CAYLEY'S* 1853 GLIDER.

"THIS ENGLISH ENGINEER IS CONSIDERED BY MANY TO BE THE FIRST SCIENTIFIC AERIAL INVESTIGATOR AND THE FIRST TO UNDERSTAND THE PRINCIPLES AND FORCES OF FLIGHT. BUT *WE* KNOW BETTER—*LEONARDO* WAS THE FIRST.

"HOWEVER, CAYLEY DID CREATE THE FIRST SUCCESSFUL GLIDER TO CARRY A PERSON, AND HE ALSO LABELED THE FOUR AERODYNAMIC FORCES OF FLIGHT—THRUST, DRAG, WEIGHT AND LIFT.

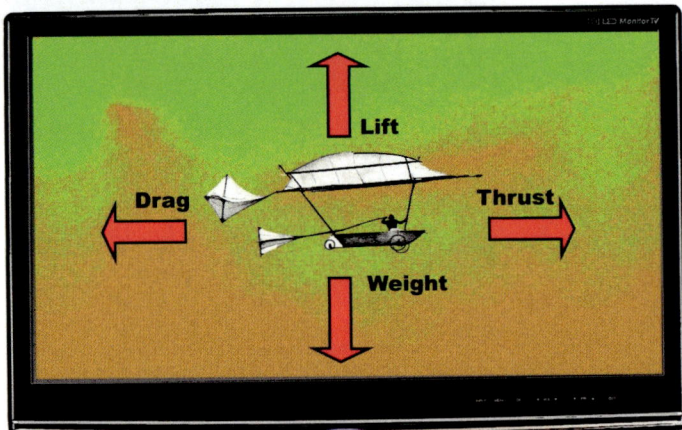

"*LIFT* IS THE UPWARD FORCE CREATED BY THE FLOW OF AIR PASSING OVER AND UNDER THE WINGS. LIFT COUNTERS THE *WEIGHT* PULLING THE AIRCRAFT DOWNWARD.

"*THRUST* IS THE FORWARD FORCE OF THE FLYING OBJECT AND *DRAG* IS THE OPPOSING MOTION THAT SLOWS IT.

"AND NOW WE HAVE 1890'S GLIDER BY *OTTO LILIENTHAL,* A HUGE HERO TO ORVILLE AND ME.

"KNOWN AS *THE GLIDER KING*, HE WAS THE FIRST PERSON TO MAKE WELL-DOCUMENTED, REPEATED, AND SUCCESSFUL GLIDING FLIGHTS.

"THE IMAGES PRINTED IN NEWSPAPERS AND MAGAZINES OF HIS FLIGHTS GARNERED INTEREST FOR MANNED FLIGHT FROM BOTH SCIENTISTS AND THE MASSES.

"OF, COURSE, HERE'S OUR OWN "WRIGHT FLYER" WE FLEW NEAR *KITTYHAWK, NORTH CAROLINA* IN 1903.

IN ADDITION TO THE STRUCTURE OF THE PLANE ITSELF, WE EVEN BUILT THE 12 HORSE-POWER *ENGINE*! ALTHOUGH OUR FIRST PLANE FLEW ONLY 12 SECONDS, WITHIN A FEW YEARS WE'D CRAFTED A PLANE THAT COULD GO ALMOST AN HOUR! WHILE IT SEEMS A BIT EGOTISTICAL TO SAY SO, I'M PROUD TO SAY THAT THIS WAS THE *FIRST-EVER MANNED AND POWERED AIRCRAFT FLIGHT!*

"HERE WE HAVE TWO WORKS BY THE RUSSIAN-AMERICAN AVIATION PIONEER OF BOTH HELICOPTERS AND FIXED-WING AIRCRAFT, *IGOR SIKORSKY*.

"HE DESIGNED AND FLEW THE WORLD'S FIRST MULTI-ENGINE FIXED-WING AIRCRAFT, *RUSSKY VITYAZ* IN 1913, AND THE FIRST AIRLINER, *ILYA MUROMETS*, IN 1914.

"*GOSSAMER CONDOR* WAS THE FIRST HUMAN-POWERED AIRCRAFT CAPABLE OF CONTROLLED AND SUSTAINED FLIGHT.

"CREATED BY *PAUL MACCREADY*, IN 1977 IT WON THE *KREMER PRIZE*, A MONETARY AWARD TO PIONEERS OF HUMAN-POWERED FLIGHT.

"*SNOWBIRD*, A HUMAN-POWERED ORNITHOPTER, WAS DESIGNED AND CONSTRUCTED BY A TEAM OF STUDENTS FROM THE *UNIVERSITY OF TORONTO*.

"ON AUGUST 2ND, 2010 IT SUSTAINED BOTH ALTITUDE AND AIRSPEED FOR 19.3 SECONDS, BECOMING THE WORLD'S FIRST SUCCESSFUL HUMAN-POWERED ORNITHOPTER.

"THE *IGOR I. SIKORSKY HUMAN POWERED HELICOPTER COMPETITION* WAS ESTABLISHED IN 1980 BY THE *AMERICAN HELICOPTER SOCIETY INTERNATIONAL* TO DEVELOP THE FIRST CONTROLLED FLIGHT OF A *HUMAN-POWERED HELICOPTER*.

"TO WIN THE COMPETITION, THE HELICOPTER WOULD HAVE TO HAVE A FLIGHT DURATION OF 60 SECONDS AND REACH AN ALTITUDE OF 3 METERS (9.8 FT)... WHILE REMAINING IN A 10 METER (32.8 FT) SQUARE!

"THE FLIGHT MUST ALSO BE CERTIFIED BY AN OFFICIAL FROM THE *FEDERATION AERONAUTIQUE INTERNATIONALE*.

"FOR MORE THAN 30 YEARS THE $250,000 PRIZE REMAINED UNCLAIMED, BUT IN JULY, 2013 IT WAS *AT LAST* WON BY *ATLAS*, CREATED BY *TODD REICHERT* AND *CAMERON ROBERTSON*!"

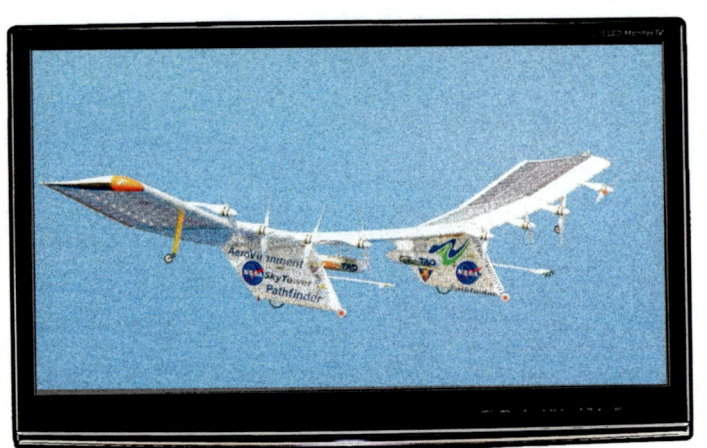

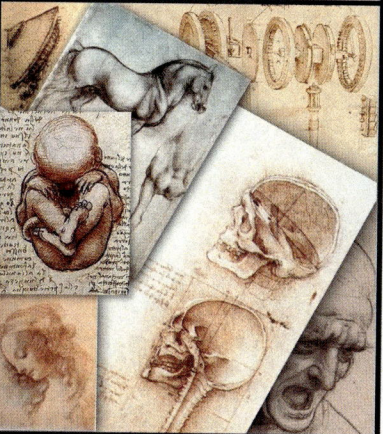

"IT'S BIG OF YOU TO BRING ME HERE TO THE SISTINE CHAPEL TO SEE THE PAINTINGS OF YOUR ADVERSARY, MICHELANGELO.

SO HE WAS YOUR ARCH NEMESIS?"

"NO, WE WEREN'T ENEMIES, BUT COMPETITORS, AS ALL WE PAINTERS WERE.

ALTHOUGH, AT ONE POINT HE DID PUBLICLY RIDICULE ME FOR NOT FINISHING ANYTHING, SUCH AS A 24-FOOT HORSE IN MILAN."

"HAHA! I CAN TELL YOU AN AMUSING STORY ABOUT US. WE WERE, HAHA, BOTH COMMISSIONED BY MACHIAVELLI, WRITER OF THE PRINCE, TO PAINT TWO MURALS IN THE SAME HALL.

HAHAHA! I'M HAPPY TO REPORT THAT NEITHER OF US FINISHED."

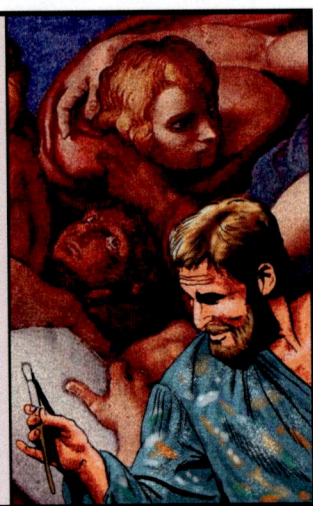

"NO QUESTION THAT MICHELANGELO WAS A FIRST-RATE PAINTER AND SCULPTOR. SURE, HIS ENORMOUS WORK ON THE SISTINE CHAPEL AND MARBLE SCULPTURE WAS TERRIFIC. BUT HE WAS ALWAYS DIRTY, COVERED WITH MARBLE DUST, WHEREAS I ALWAYS WAS ELEGANTLY DRESSED!

"WE WERE BOTH BROUGHT UP IN SIMILAR VILLAGES OUTSIDE OF FLORENCE, FROM MIDDLE CLASS FAMILIES. BUT HE WAS A FAMILY MAN, AND I WAS A LONER.

"I LIVED TO BE 67, WHEREAS THAT RASCAL LIVED TO BE 90!"

AND SPEAKING OF FLORENCE...

"WE CAN NOW DO IN THE BLINK OF AN EYE WHAT USED TO TAKE THREE MONTHS BY HORSE-DRIVEN CARAVAN..."

"BEHOLD! THE "WOMB" OF THE RENAISSANCE-- FLORENCE!"

220

"I WAS BORN IN 1452, AND THE 150 YEARS PRIOR TO MY BIRTH SAW HUNDREDS OF "LOST" ANCIENT GREEK AND ROMAN MANUSCRIPTS FLOODING INTO FLORENCE FROM EUROPEAN MONASTERIES AND FROM ASIA, THE BOUNTY OF TRAVELING MERCHANTS AND COLLECTORS.

AFICIONADOS OF GRAECO-ROMAN LITERATURE WERE ADHERENTS TO AN INTELLECTUAL MOVEMENT WHICH SOUGHT TO IMPROVE MANKIND THROUGH STUDY OF THE ANCIENTS-- HUMANISM.

"THE HUB AND FOUNDER OF HUMANISM WAS THE COLLECTOR AND POET, FRANCESCO PETRARCA-- AKA PETRARCH.

HE COINED THE NOW-FAMOUS HISTORICAL TERM "THE DARK AGES" TO DESCRIBE WHAT HE PERCEIVED TO BE THE INTELLECTUAL STAGNATION AND DECLINE THAT FOLLOWED THE FALL OF THE ROMAN EMPIRE IN 476 CE.

"COLUCCIO SALUTATI, CHANCELLOR OF FLORENCE, OWNED THE LARGEST LIBRARY IN FLORENCE-- OVER 800 BOOKS!

"BASING HIS BRILLIANT ORATORY AND WRITING ON SUCH CLASSIC AUTHORS AS VIRGIL AND CICERO (WHOSE WORK PETRARCH HAD REDISCOVERED), HE HELPED PERSUADE FLORENCE'S ELITE OF THE MERIT OF THE HUMANIST CAUSE.

"THE POET GIOVANNI BOCCACCIO, WRITER OF THE DECAMERON, FOUND TWO WORKS BY THE ROMAN HISTORIAN TACITUS, PRESERVING IMPORTANT CHRONICLES OF THE HISTORY OF THE ROMAN EMPIRE.

"POGGIO BRACCIOLINI, AMONG OTHER WORKS ON ASTRONOMY, MINING, AND AGRICULTURE, FOUND WHAT TURNED OUT TO BE QUITE INFLUENTIAL TO THE DEVELOPMENT OF SCIENCE: DE RERUM NATURA-- ("ON THE NATURE OF THINGS") BY LUCRETIUS. THIS DIDACTIC POEM EXPLAINED THE UNIVERSE VIA NATURAL, ATOMISTIC MEANS, RATHER THAN BY SUPERNATURAL INTERVENTION.

HE ALSO DISCOVERED A WORK THAT WAS PARTICULARLY INFLUENTIAL ON ME: DE ARCHITECTURA LIBRI DECEM-- ("THE TEN BOOKS OF ARCHITCTURE")-- BY THE ROMAN ARCHITECT AND ENGINEER, VITRUVIUS."

"WELL, I JUMPED AHEAD OF MYSELF A LITTLE THERE. BACK TO FLORENCE!

SO, WHEN I FIRST CAME HERE AT AGE 15 IT WAS WONDERFUL--I'D GROWN UP IN THE SMALL HILL TOWN OF VINCI, SO THE COSMOPOLITANISM OF FLORENCE WAS JUST ASTOUNDING.

I APPRENTICED IN THE STUDIO OF VERROCCHIO, WHERE I LEARNED ALL THE NECESSARY SKILLS I'D NEED TO BE A WORKING ARTIST, SUCH AS HOW TO WELD, HOW TO CREATE FRAMES FOR PAINTINGS, HOW TO CAST IN BRONZE, OR HOW TO CREATE PAINTS FROM SCRATCH.

"MASTER VERROCCHIO ONCE SCULPTED A STATUE OF DAVID AND GOLIATH, AND POPULAR WISDOM HOLDS THAT I MODELED FOR IT."

SO TELL ME, THEN-- IS IT AN ACTUAL SCULPTURE OF YOU?

HMM... LET'S JUST SAY THAT I'D BE PLEASED TO BE THOUGHT OF AS THE MODEL, FOR MY MASTER WAS BRILLIANT!

"HOWEVER, THOUGH VERROCCHIO WAS INDEED A MASTER, WHEN I WAS HIS APPRENTICE HE ONCE ASKED ME TO PAINT THE *ANGEL* ON THE FAR LEFT SIDE OF THIS IMAGE, AND MUCH OF THE *BACKGROUND* SCENERY...

WITH A MIXTURE OF PRIDE AND REMORSE I TELL YOU THAT HE WAS SO PLEASED AND ASTONISHED WITH MY SKILL THAT HE *NEVER* PAINTED AGAIN!"

BUDDHIST *MONKS* IN *TIBET* HAVE A SAYING--"IF THE STUDENT DOES NOT SURPASS THE MASTER THEN THE MASTER IS NO GOOD."

SOUNDS ABOUT RIGHT.

"WELL, OUR TRIP TOGETHER WOULDN'T BE COMPLETE WITHOUT ME BRINGING YOU TO THE *SANTA MARIA DELLE GRAZIE* MONASTERY HERE IN *MILAN*..."

225

THE END

"I AM AWARE OF SOME OF THE TRAGIC REPERCUSSIONS OF THE *CHEMICAL FIGHT AGAINST INSECTS* TAKING PLACE IN FRANCE AND ELSEWHERE, AND I DEPLORE THEM..."

"MODERN MAN NO LONGER KNOWS HOW TO FORESEE AND TO FORESTALL. HE WILL END BY *DESTROYING THE EARTH* FROM WHICH HE AND OTHER LIVING CREATURES DRAW THEIR FOOD..."

"POOR BEES, POOR BIRDS, POOR MEN..."

THESE ARE THE WORDS AND IMAGE OF *DR. ALBERT SCHWEITZER.*

HE WROTE THIS DIRE WARNING IN A 1956 LETTER TO A FRENCH BEEKEEPER AFTER LEARNING THAT THE APIARIST'S *HONEYBEES* WERE BEING *KILLED* BY INDISCRIMINATE SPRAYING OF *INSECTICIDES.*

SIX YEARS LATER THE SCHWEITZER MEDAL WAS AWARDED BY THE ANIMAL WELFARE INSTITUTE TO MARINE BIOLOGIST *RACHEL CARSON.*

CARSON BECAME KNOWN AS ONE THE FIRST *TRUE ECOLOGISTS* IN THE MODERN SENSE, AFTER PUBLICATION OF HER SEMINAL WORK, *SILENT SPRING.*

HAHA! LOOK AT *DADDY* IN THAT *BEEKEEPER* OUTFIT. BUT WHY'S HE WEARING IT?

AH, YOUR FATHER'S DULCET TONES ARE MAKING ME *SLEEPY* ALREADY.

Carson's first book, *THE SEA AROUND US*, stated that human garbage, which included a variety of dreadful chemicals, could be found from pole to pole.

Though this book was a bestseller, it was with *SILENT SPRING* that Carson single-handedly did more to save the environment from runaway chemical pollution than any other book or any other group of environmental activists, before or since.

She can truly be considered the mother of the "GREEN REVOLUTION".

"Let's back up a bit, to the post-World War II **1940s**. During this period, hundreds of new chemical products appeared on the market. Americans were inundated by such snappy slogans as..."

DU PONT
BETTER THINGS FOR BETTER LIVING ...THROUGH CHEMISTRY

One of the most important new types of chemical product was PLASTICS--

"THERE'S A GREAT FUTURE IN PLASTICS."

THE GRADUATE. You're *NEVER* going to stump me on movie quotes, Cecil.

To the American public these pesticides were a true *GODSEND* capable of annihilating all the nasty organisms threatening crops and health.

—The other was PESTICIDES.

PESTROY
TRADE MARK REG. U.S. PAT. OFF.
DDT POWDER
LONG-LASTING!
DOG'S best friend!
Kills fleas!
39¢ 3 OZ. ACTIVATED POWDER
PRODUCT OF *SHERWIN-WILLIAMS* RESEARCH

"The edge of the sea is a strange and beautiful place."

Rachel Carson

"I was born in 1907 and grew up on a small farm in Pennsylvania, the youngest of three kids.

"I was ever *outdoors*, exploring nature's wonders. Mother encouraged a deep *respect* for the natural world, to revel in its beauty and mystery.

"A former *schoolteacher*, Mother also inspired my love for learning, reading, and writing.

"I grew up wanting to be a writer. I was only *eleven* when my first published story, a World War One tale of *"A Battle in the Clouds"* between German and Canadian aviators, appeared in the pages of *St. Nicholas Magazine!*

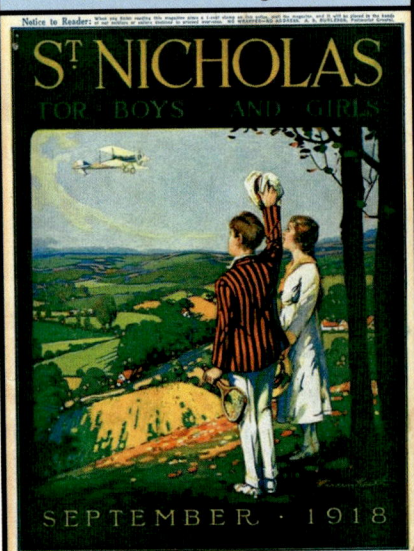

"In school I was a *dedicated student*, eventually earning the necessary *scholarships* that paid my way to college.

"Mother helped earn my tuition, too, by giving piano lessons, doing odd jobs, and even selling the *family china!* Yes, sir, she certainly sacrificed a lot for me.

"I went to *Pennsylvania College for Women* (now Chatham University,) and, no surprise, I started out as an English major.

"But while taking a required biology class I became re-enamored with the study of nature and was encouraged by *Professor Mary Skinker*, my instructor and mentor, to switch my major to *zoology*.

Professor Skinker had a *profound* influence on me, and I eventually followed her to *Johns Hopkins University* for grad school, where she went to finish her PhD.

"I graduated magna cum laude from PCW, earning a *full scholarship* to Johns Hopkins University.

"In the summer of 1929, after graduating from PCW, I first went to *Woods Hole Marine Biological Laboratory* in Massachusetts to study marine biology, and finally first laid eyes on the *ocean!*

"It was *love at first sight!* I spent a *lot* of time at this fabulous research center, both before and after I earned my master's in *marine zoology* from Johns Hopkins.

"I understand that a small *statue* of me has even been installed at Woods Hole, commemorating my time there. *Very* flattering.

"Years after I left, I do believe that *Igor Gamow* also spent a couple of summers taking courses there. I'm sure that he and I share a mutual love and inspiration for the place.

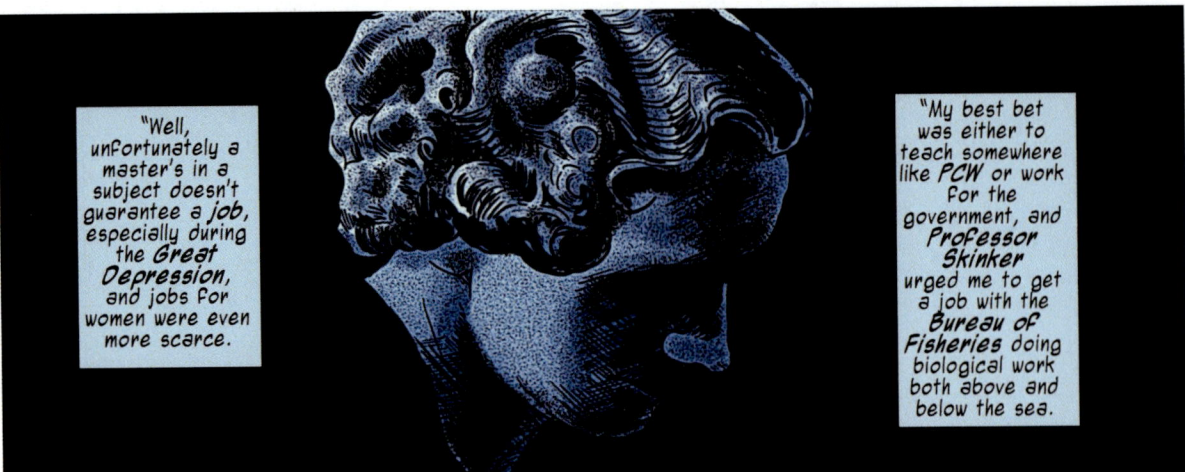

"Well, unfortunately a master's in a subject doesn't guarantee a *job*, especially during the *Great Depression*, and jobs for women were even more scarce."

"My best bet was either to teach somewhere like *PCW* or work for the government, and *Professor Skinker* urged me to get a job with the *Bureau of Fisheries* doing biological work both above and below the sea."

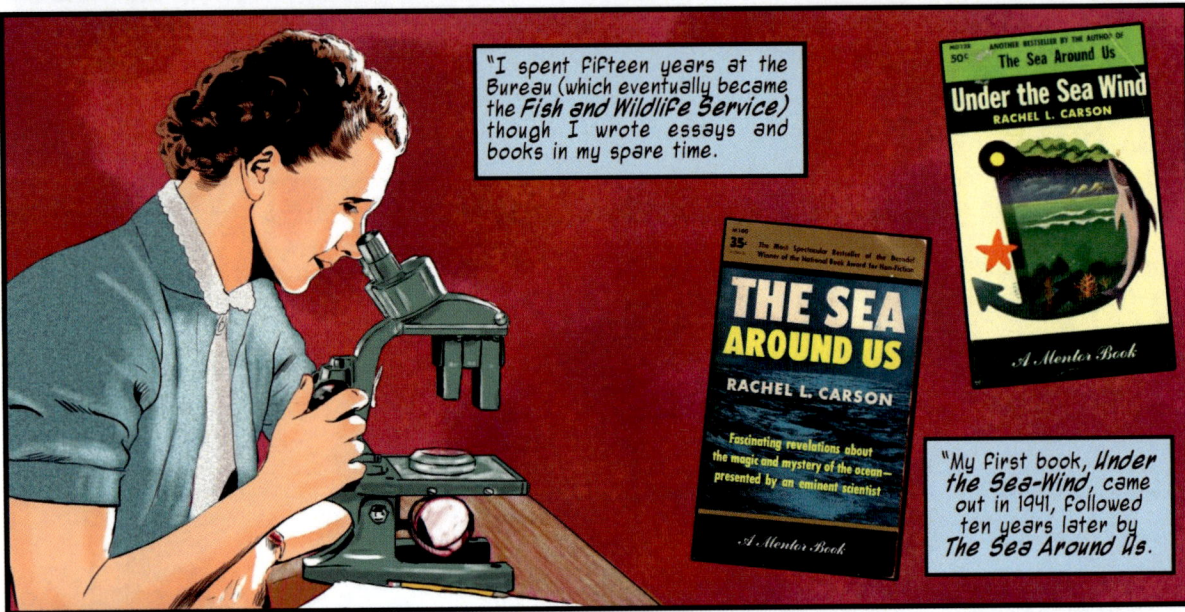

"I spent fifteen years at the Bureau (which eventually became the *Fish and Wildlife Service*) though I wrote essays and books in my spare time."

"My first book, *Under the Sea-Wind*, came out in 1941, followed ten years later by *The Sea Around Us*."

"I never fully enjoyed my job--too much *bureaucracy*! But I finally was able to retire in 1952 (fairly early, luckily) to devote all my time to writing."

"1955 saw the final part of my 'sea trilogy' published-- *Edge of the Sea.*"

"And here we stand at the edge of the sea, on this beach... Mr. Tompkins, how could *anyone* gaze out at that ocean and not be *captivated* by it?"

"Any who study it are forever changed, as I was."

"Water covers some 71% of the Earth's surface, so there's plenty to study."

"Fifty percent of sea water is at depths of more than 1000 feet, and we have almost no clue as to what occurs down there."

"We actually know more about *other planets* than we know about Earth's deep oceans, which are devoid of light and mostly unexplored."

"Our *ignorance* is as deep as the ocean!"

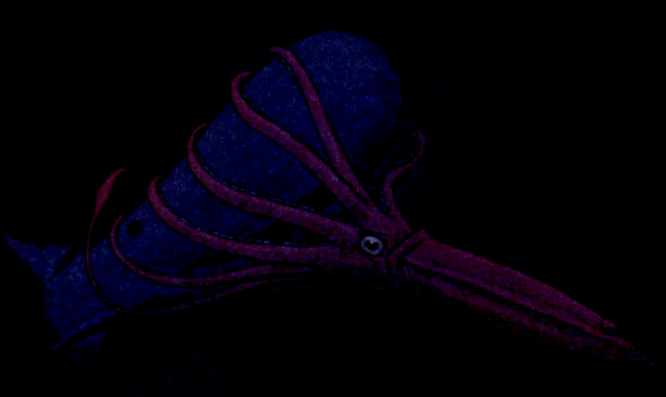

"We do have some *hints* as to what's happening down there, though.

We've found dead *sperm whales* marred with platter-sized *sucker marks*..."

"Perhaps the aftermath of a *battle* between the whale and a *giant squid*? Taking place half a mile beneath the sea in total darkness? The idea just *blows my mind!*"

"Also, we ask, 'Did *sea monsters* ever exist?' and if so, 'Do they *still* exist?' The answer may be... '*Yes!*'"

"In 2013 the 18-foot-long carcass of a "sea serpent" drifted into the shallow waters of a California beach."

"This mysterious creature, an *oar fish*, is rarely seen by humans due to the fact that it lives in depths of 3000 feet below the water surface."

The 400-pound creature required *14 people* to drag it ashore!"

SEA MONSTER, INDEED! SO, IT'S THE SEA'S MYSTERY THAT ATTRACTS YOU TO IT?

Well, when I was a child I loved reading *maritime adventure stories*...

"...books such as Herman Melville's *Moby Dick*, Robert Louis Stevenson's *Treasure Island*, and Jules Verne's *20,000 Leagues Under the Sea*. Ah, that mysterious Indian prince *Captain Nemo* and his amazing submarine, the *Nautilus!*"

"I can't help but compare Nemo with modern-day scientists such as *William Beebe* and his deep sea *Bathysphere*...

"...Or *Jacques-Yves Cousteau*. I loved his books and films, and his marine biology-lab-boat, the *Calypso*.

We speak of a 'man for all seasons' but here was a '*man for all seas*'-- naval officer, explorer, conservationist, filmmaker, inventor, scientist, photographer, author and researcher.

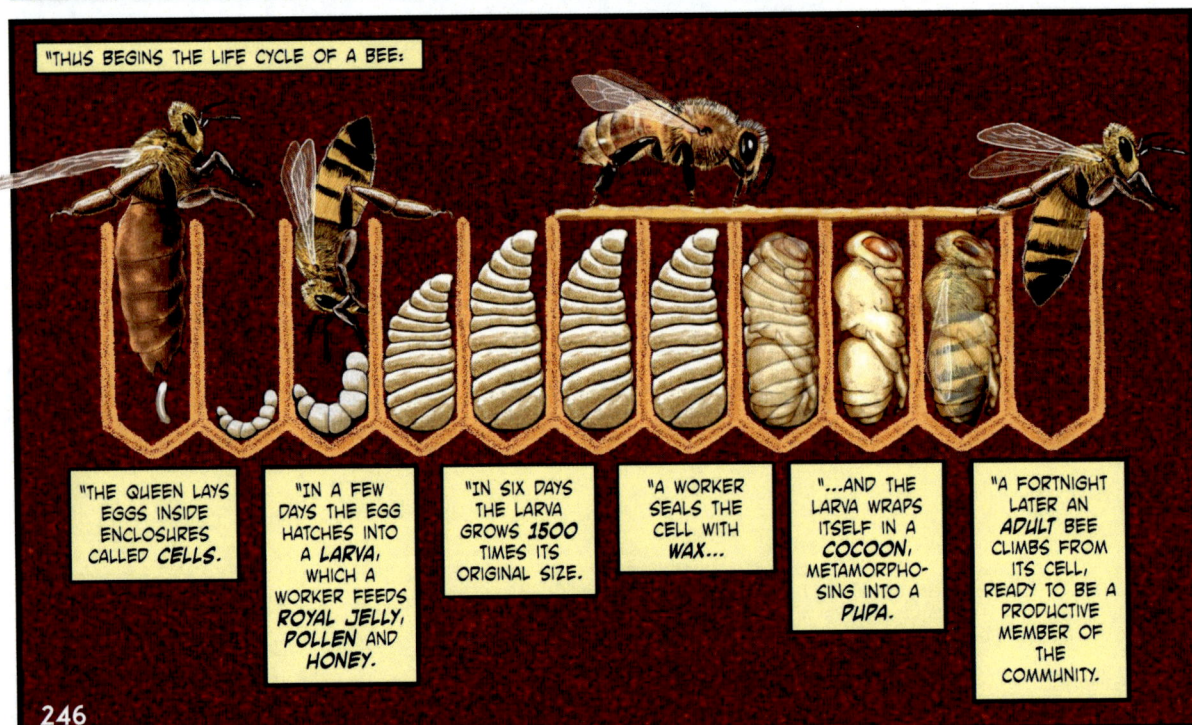

"WELL, WITH ALL THOSE BEES BEING BORN, DOESN'T THE HIVE EVENTUALLY GET **OVERPOPULATED**?"

"IT DOES. WHEN THIS HAPPENS THE OLD QUEEN, NOW REPLACED BY A NEW ONE, TAKES A **SWARM** OF ABOUT HALF THE POPULATION WITH HER TO ESTABLISH A NEW COLONY.

Karl, tell him about your discovery, the *waggle dance.*

OH, I **KNOW** THE WAGGLE DANCE!

BUT SERIOUSLY. TELL ME ABOUT THE WAGGLE DANCE.

HONEYBEES, I DISCOVERED, HAVE A VERY **SOPHISTICATED** METHOD OF COMMUNICATION.

WHILE NOT EXACTLY A "LANGUAGE", AS I FIRST CALLED IT (SINCE IT ISN'T SPOKEN, SUNG, OR WRITTEN), IT **IS COMMUNICATION**... A COMMUNICATION OF **MOTION**.

""BEFORE I TELL YOU ABOUT THE WAGGLE DANCE, LET ME TELL YOU HOW HONEYBEES SEE.

NOT ONLY DO THEY HAVE **COLOR VISION** (THOUGH NOT AS REFINED AS YOURS OR MINE), BUT THEY HAVE **FIVE** EYES AND CAN SEE **ULTRAVIOLET** AND **POLARIZED** LIGHT.

THUS, THEY'RE **ALWAYS AWARE** OF THE SUN'S LOCATION, EVEN IF THE SUN ISN'T VISIBLE TO OUR EYES, SUCH AS ON A CLOUDY DAY.

AS SUNLIGHT CUTS THROUGH THE ATMOSPHERE IT'S POLARIZED IN THE SUN'S DIRECTION AS VIEWED FROM EARTH, AND A HONEY BEE IS AWARE OF THE SUN'S DIRECTION EVEN IF FACING AWAY FROM IT.

"Simple" Eyes

Compound Eyes

"AN INTERNAL **BIOLOGICAL CLOCK** ALSO ALLOWS THE BEE TO ESTIMATE THE LOCATION OF THE **SUN** IN ITS PATH ACROSS THE SKY EVEN IF THE BEE'S BEEN INSIDE A DARK HIVE FOR HOURS.

"EVEN IN THE DARK "UP" AND "DOWN" ARE CLEARLY DEMARKED BECAUSE **GRAVITY** IS THE COMMON REFERENCE POINT AND THUS EQUALS "DOWN", WITH "UP" AS A PROXY FOR THE SUN.

"A 'SCOUT' WORKER BEE RETURNED FROM A FORAGING MISSION PERFORMS A CHARACTERISTIC **FIGURE-EIGHT** MOVEMENT IN ORDER TO COMMUNICATE TO HER HIVE THE **DIRECTION** AND **DISTANCE** OF NECTAR- AND POLLEN-RICH FLOWERS, WATER SOURCES, OR POTENTIAL NEW LOCATIONS TO MOVE THE COLONY.

"THE **DIRECTION** OF THE ZIG-ZAG 'WAGGLE' PART OF THE DANCE COMPARED TO THE "UP" POSITION IS THE **SAME ANGLE** THE OTHER BEES SHOULD FLY AWAY FROM THE SUN TO REACH THE DESIRED LOCATION.

"IF THE BEE WAGGLES DIRECTLY TOWARDS "UP", THEN THE BEES KNOW TO FLY STRAIGHT TOWARDS THE SUN, AND VICE VERSA.

"ALSO, THE **LENGTH OF TIME** THE WORKER WAGGLES INDICATES THE **DISTANCE** OF THE LOCATION-- ONE SECOND FOR EACH KILOMETER OF FLIGHT."

SORRY, TOMPKINS. I KNOW YOU **HATE SHRINKING**, BUT IT'S THE ONLY WAY WE'RE ABLE TO ACCOMPLISH WHAT WE'RE ABOUT TO DO.

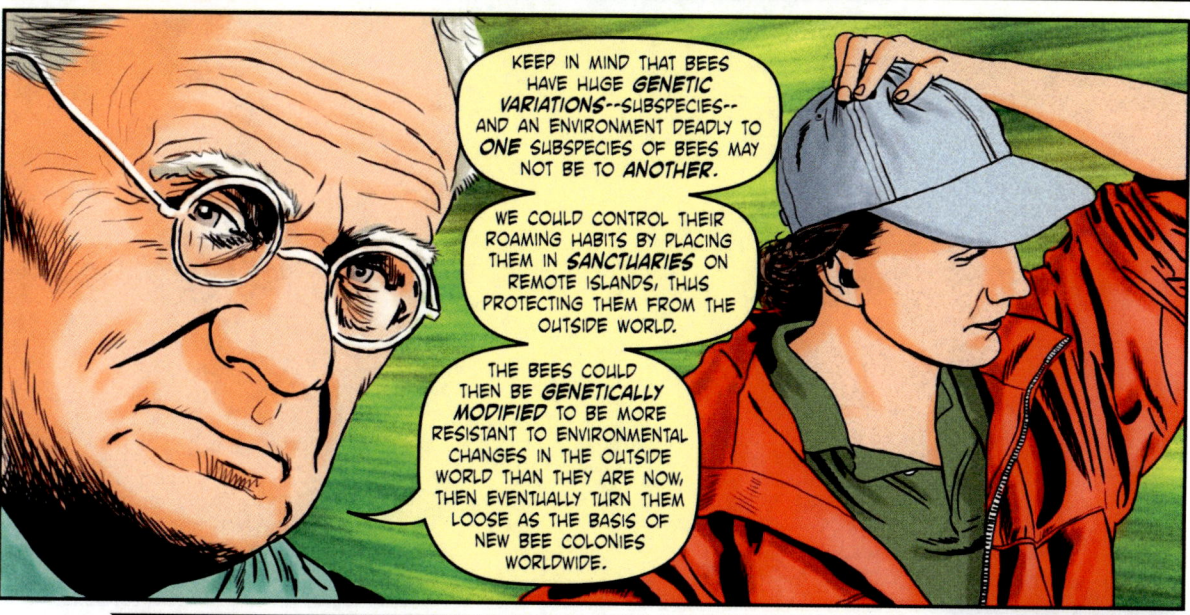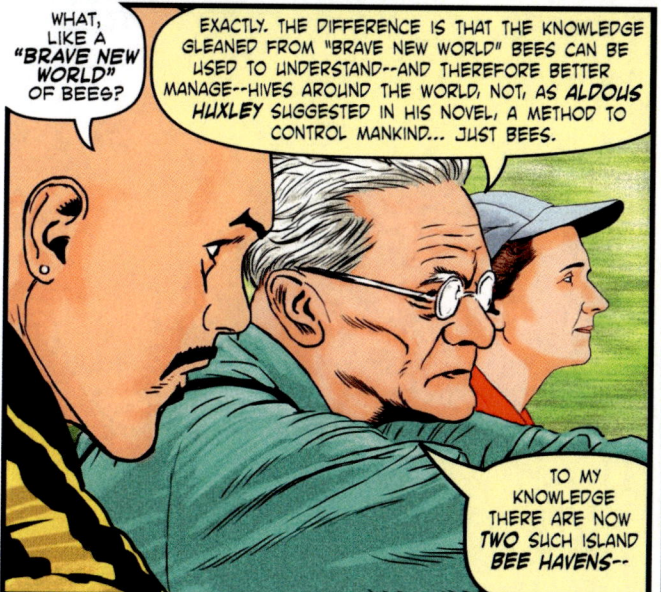

"AHHH... HIVE, SWEET HIVE!"

"NOW, MELISSA REGURGITATES THE NECTAR INTO THE MOUTH OF A WAITING HOUSEHOLD WORKER..."

"...WHO DUMPS THE NECTAR INTO AN EMPTY CELL OF THE HONEYCOMB. SHE AND OTHER HOUSEHOLD BEES FAN THEIR WINGS OVER IT TO EVAPORATE THE EXCESS MOISTURE."

"ABOUT FIVE DAYS LATER WE HAVE HONEY, WHICH THE BEES SEAL OVER WITH WAX FOR PRESERVATION."

"THE WAX IS SECRETED FROM SPECIAL GLANDS, AND IS ALSO USED TO BUILD THE COMB."

"THEY HAVE TO EAT A POUND OF HONEY JUST TO PRODUCE AN OUNCE OF WAX, AND--"

GRAHHH!

"OH! WHAT'S THAT SOUND??"

"WE'D BETTER RUN OUTSIDE AND SEE WHAT THE UPROAR IS!"

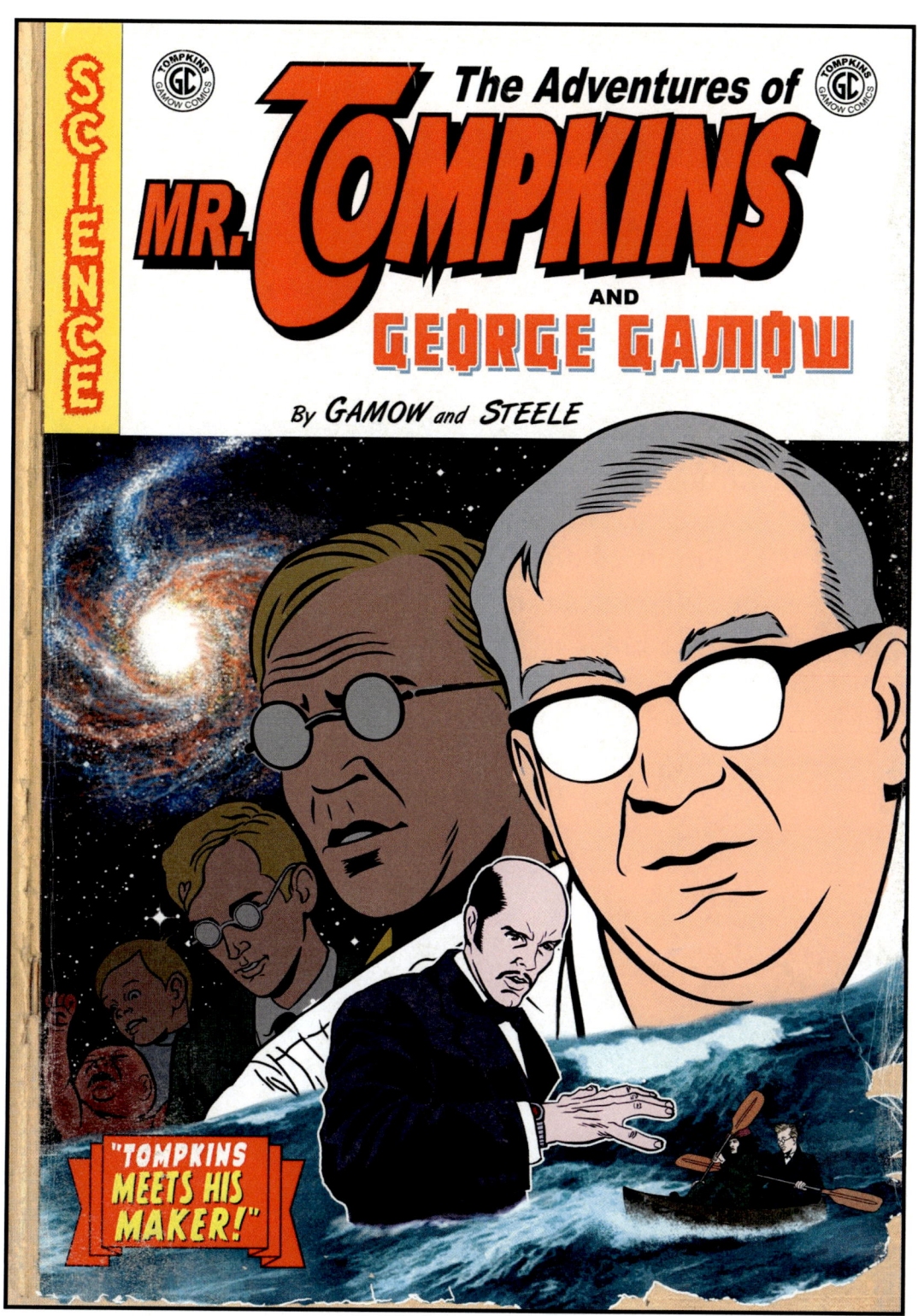

"Although the language was very technical, and some of the passages did not make sense to him at all, he still felt that he was learning something."

George Gamow
Mr. Tompkins Learns the Facts of Life

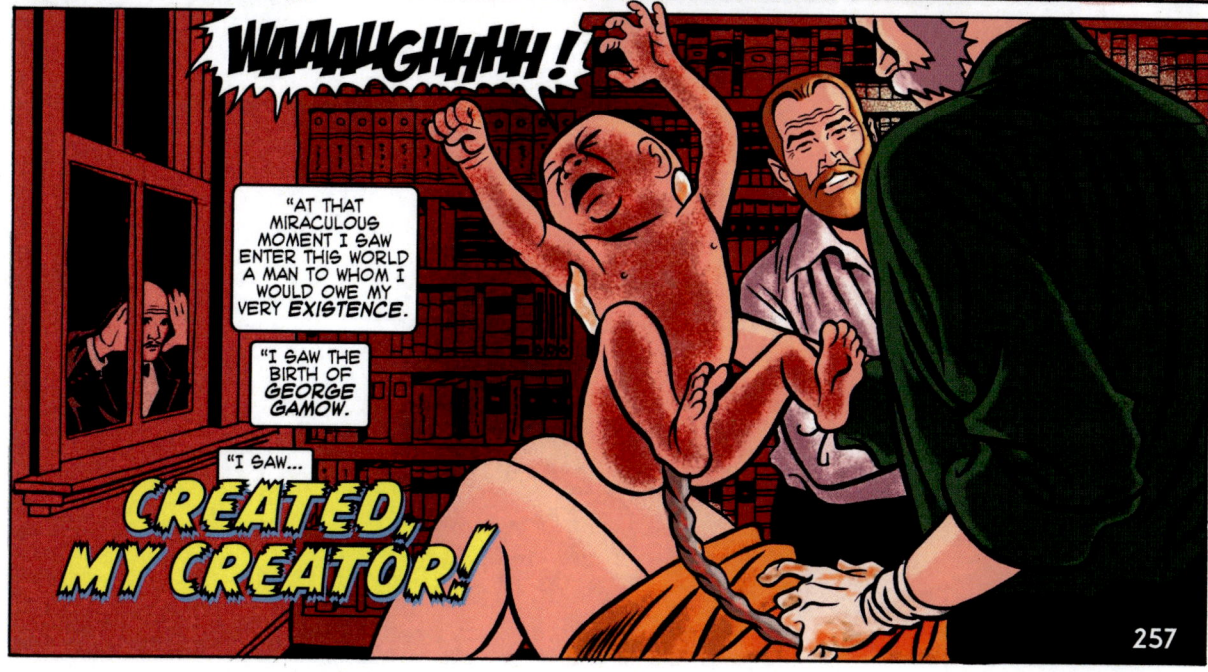

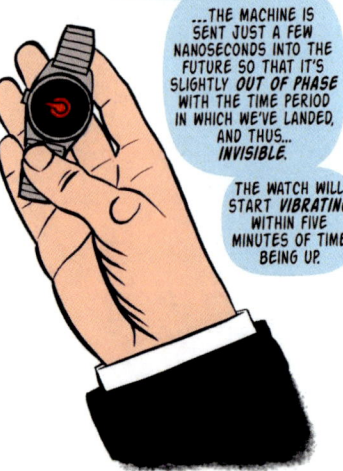

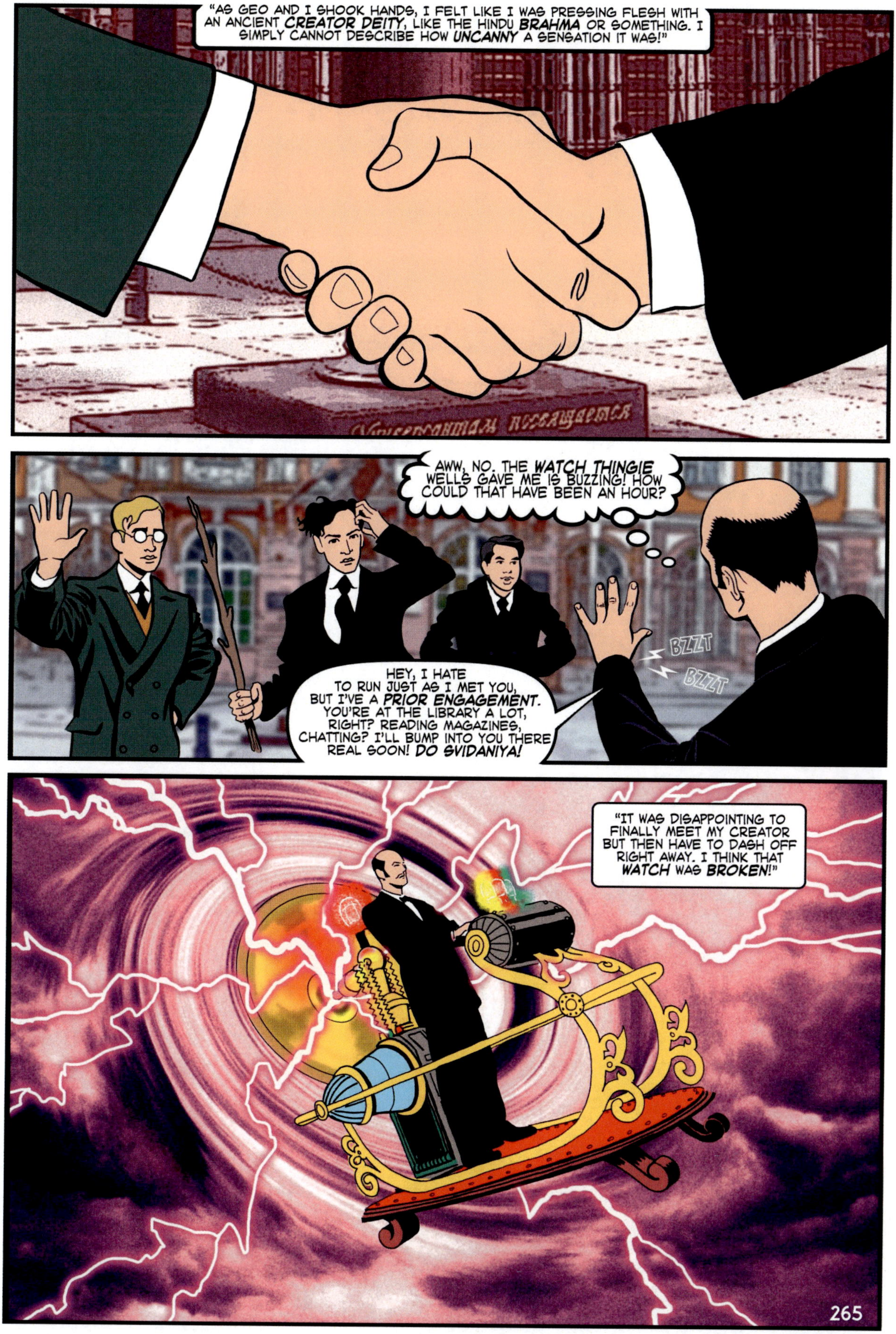

"AND THERE YOU HAVE IT."

"SO FAR THE CONGRESS HAS BEEN PRETTY UNEVENTFUL, EXCEPT WHEN EVERYONE DEMANDED THAT I--*HA, HA!*--STOP SPEAKING FRENCH AND SWITCH TO ENGLISH."

"BEFORE THE CONGRESS IS OVER I NEED TO TALK TO *JAMES CHADWICK* OVER THERE ABOUT HIS DISCOVERY OF THE *NEUTRON*--"

"THAT SUBATOMIC PARTICLE *WITHOUT* AN ELECTRIC CHARGE."

"PRECISELY. THE MOST IMPORTANT OUTCOME OF THIS CONFERENCE FOR ME, THOUGH, IS THAT I DO NOT HAVE TO RETURN TO RUSSIA!"

"AT DINNER A COUPLE OF NIGHTS AGO I TOLD *MADAME CURIE* HERE ABOUT HOW I DID NOT WANT TO GO BACK, BUT WAS AT MY WIT'S END AS TO WHAT TO DO."

"SHE TALKED TO *PAUL LANGEVIN*, CHAIRMAN OF THIS CONGRESS-- AND *ALSO* A MEMBER OF THE *FRENCH COMMUNIST PARTY*--AND THEY SOMEHOW FINAGLED A WAY FOR ME TO STAY HERE IN PARIS!"

"GEO, THAT'S GREAT!"

BZZT BZZT

"UM, GEO? I APOLOGIZE, BUT, UH... I REALLY NEED TO USE THE FACILITIES. I'LL BE RIGHT BACK!"

"ERM...ALRIGHT, TIOMKIN. I SWEAR, YOU ARE LIKE THE WHITE RABBIT IN *ALICE IN WONDERLAND*."

277

"YEP. I ACTUALLY LEARNED YOU WERE MY CREATOR IN A DREAM.

"ALBERT EINSTEIN TOLD ME. IT WAS A BIT OF A SHOCK, REALLY...

WHAT IS THE MATTER?

I'M NOT A REAL PERSON? JUST... MADE UP?

YES, LAD, YOU'RE ONLY A FICTIONAL CHARACTER.

BUT TAKE DR. EINSTEIN'S ADVICE AND YOU'LL BE FINE--

WISH UPON A STAR AND CALL ME IN THE MORNING!

BUT... I'VE GOTTEN OVER IT PRETTY WELL, REALLY.

SINCE I DREAMED ABOUT IT IT JUST DOESN'T SEEM QUITE "REAL".

PERHAPS A CHARACTER IN A STORY ACCEPTS THINGS AT A FASTER RATE JUST TO FURTHER THE PLOT.

AND, WELL, MAYBE SINCE I'M A FICTIONAL CHARACTER I TAKE A LOT MORE "IMPOSSIBLE" THINGS IN STRIDE.

I'D NEVER LEARN ANYTHING, NOR WOULD THE READER.

THERE JUST REALLY ISN'T ROOM IN A SHORT STORY OF LIMITED LENGTH TO LINGER ON ANY ANGST I MIGHT FEEL AT LEARNING I'M NOT "REAL".

AND SINCE YOU AND I ARE INTERACTING, MIGHT YOU BE JUST A FICTIONAL CHARACTER, TOO?

BUT ANYWAY... ALBERT DIDN'T TELL ME MUCH MORE THAN THAT YOU CREATED ME. SO FILL ME IN, IF YOU DON'T MIND.

WHEN I CAME TO THE UNITED STATES IN 1934 TO ACCEPT THE PROFESSORSHIP AT GEORGE WASHINGTON UNIVERSITY I MET A MAN NAMED TOMPKINS.

THAT "KINS" PART OF THE NAME JUST STRIKES ME AS FUNNY! HEH. I DON'T KNOW WHY!

"OF COURSE, TOMPKINS REMINDED ME OF "TIOMKIN", THE NAME YOU GAVE ME. SO, I COMBINED HIM WITH YOUR NAME AND APPEARANCE WHEN, IN 1938, I WROTE A SERIALIZED STORY FOR DISCOVERY MAGAZINE ABOUT A BANK CLERK WHO WAS INTERESTED IN SCIENCE. THUS WAS BORN YOU, MR. TOMPKINS!

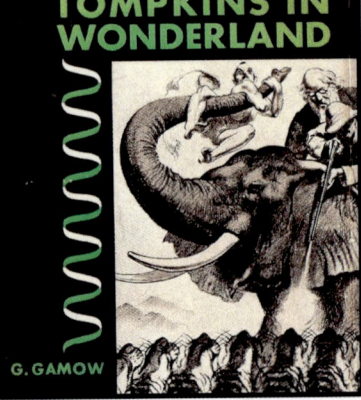

"LAST YEAR A COLLECTED VERSION OF THE STORIES WAS RELEASED-- MR. TOMPKINS IN WONDERLAND. IT'S SELLING PRETTY WELL, I MUST SAY.

BY THE WAY-- HOW DID YOU FIND ME AT HOME? HOW HAVE YOU FOUND ME ALL THESE TIMES THROUGHOUT MY LIFE?

DOES THE NAME H.G. WELLS MEAN ANYTHING TO YOU?

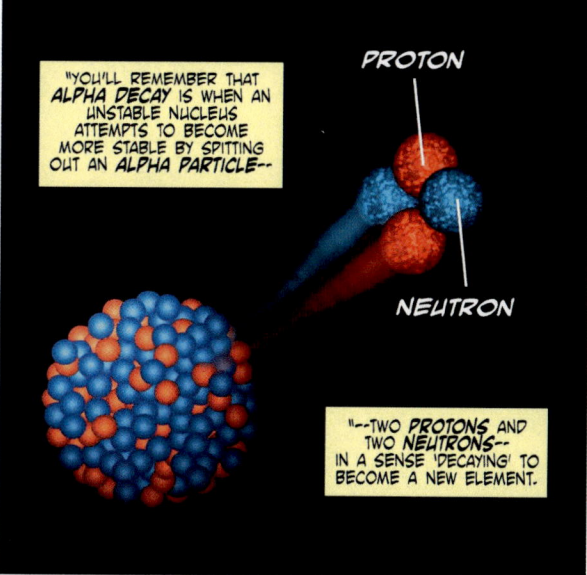
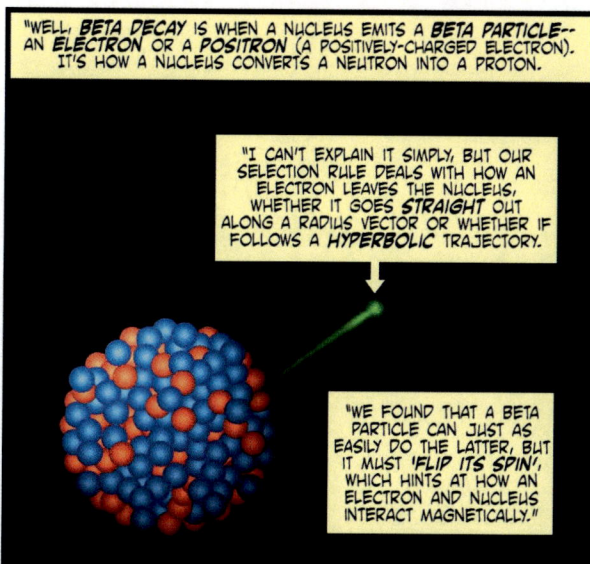

"IN 1938 TELLER AND I ORGANIZED OUR FOURTH **WASHINGTON CONFERENCE ON THEORETICAL PHYSICS**, AND DEDICATED IT TO THE SUBJECT OF NUCLEAR PROCESSES IN STARS.

"AMONG THE MANY LUMINARIES WHO ATTENDED WAS **HANS BETHE**, A NUCLEAR PHYSICIST FROM CORNELL UNIVERSITY. WHEN HE ARRIVED HE KNEW **NOTHING** OF A STAR'S INNARDS BUT BY THE END OF THE SUMMIT HE HAD AN EXPLANATION FOR NUCLEAR REACTIONS INVOLVING HYDROGEN AND CARBON! A WEEK LATER, ON HIS WAY BACK TO CORNELL, HE COULD CALCULATE THE SUN'S OBSERVED **LUMINOSITY** (THE TOTAL AMOUNT OF ENERGY IT EMITS)! **A BRILLIANT MAN!**

"ABOUT A YEAR AGO I TOOK THE FAMILY ON VACATION TO **RIO DE JANEIRO**, AND AT THE **CASSINO DA URCA** I MET A YOUNG THEORETICAL PHYSICIST NAMED **MARIO SCHOENBERG**.

"WE STRUCK UP A FRIENDSHIP AND I ARRANGED A **GUGGENHEIM FELLOWSHIP** FOR HIM IN WASHINGTON SO WE COULD WORK ON NUCLEAR ASTROPHYSICS TOGETHER.

GEO, I DO BELIEVE THAT **ROULETTE** AND **STARS** MAY HAVE SOMETHING IN COMMON.

OH? HOW'S THAT, MARIO?

DEAR, LET'S BET THREE CHIPS ON 14.

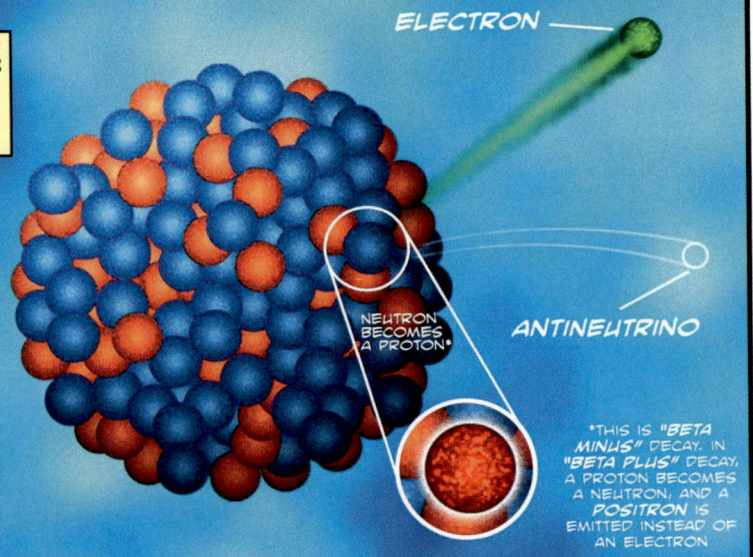

"EVERYTHING WORKED OUT **GREAT!** WE LEARNED A LOT ABOUT A STAR'S FINAL DAYS, AND ONE PROCESS BY WHICH IT MIGHT **EXPLODE**--WITHIN THE STELLAR INTERIOR, ATOMIC NUCLEI LOSE ENERGY BY ABSORBING AN ELECTRON AND RE-EMITTING A BETA PARTICLE WITH A **NEUTRINO-ANTINEUTRINO** PAIR.

"**NEUTRINOS** ARE NEUTRAL PARTICLES PRODUCED IN BETA DECAY AND HAVE ALMOST NO INTERACTION WITH MATTER. **ANTINEUTRINOS** ARE PRODUCED WHEN A NEUTRON TURNS INTO A PROTON DURING BETA DECAY.

"NEUTRINOS AND ANTINEUTRINOS HAVE SUCH POWER AND ENERGY THAT THEY'RE ABLE TO PENETRATE A STAR LIKE A **SWARM OF MOSQUITOES** THROUGH CHICKEN WIRE, SUCKING POWER AWAY FROM THE STAR AND RAPIDLY COOLING IT.

ELECTRON

NEUTRON BECOMES A PROTON*

ANTINEUTRINO

*THIS IS "**BETA MINUS**" DECAY. IN "**BETA PLUS**" DECAY, A PROTON BECOMES A NEUTRON, AND A **POSITRON** IS EMITTED INSTEAD OF AN ELECTRON.

"THE INTERNAL PRESSURE DROPS, THE STAR **COLLAPSES**, AND A FIERY EXPLOSION--A **SUPERNOVA**--EMITS AS MUCH ENERGY AS OUR SUN IS EXPECTED TO YIELD DURING ITS ENTIRE **LIFESPAN**...

"...EJECTING ALL OR MOST OF THE STAR'S MATERIAL AT A SPEED OF UP TO **30,000 KM/S**--A TENTH OF THE **SPEED OF LIGHT!**

"MARIO AND I CALLED THIS THE "**URCA PROCESS**" TO ANALOGIZE A STAR'S RAPID HEAT LOSS WITH A GAMBLER'S LOSS OF MONEY AT THE **ROULETTE** TABLE.

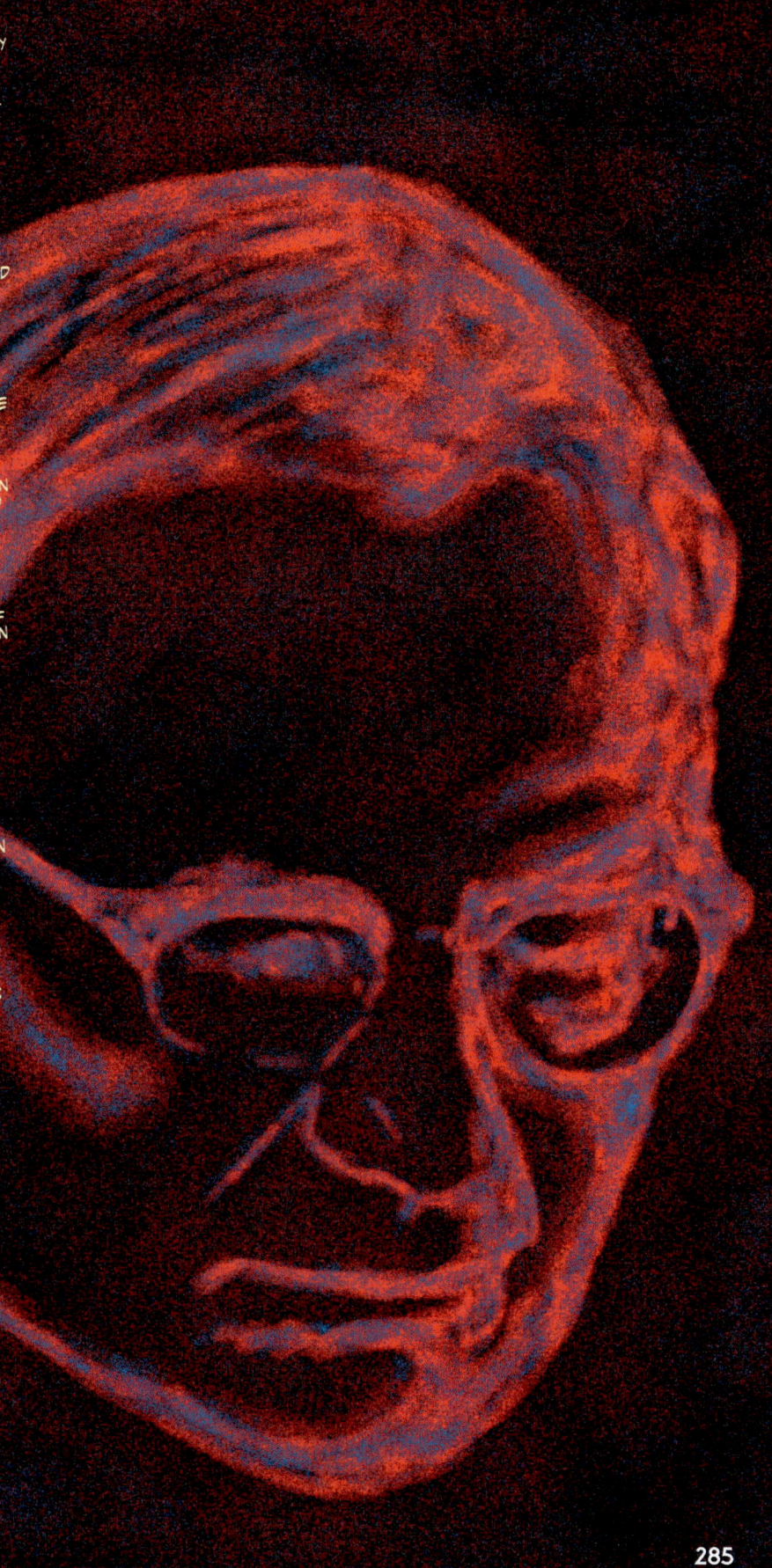

"NOW HERE'S WHERE I COME IN. BY THE 1940s WE BELIEVED THAT THE UNIVERSE IS CHEMICALLY *HOMOGENOUS*, AND THAT THE RELATIVE ABUNDANCE OF DIFFERENT ELEMENTS IS FAIRLY WELL-REPRESENTED BY THE CONSTITUTION OF THE SUN, STARS, AND INTERSTELLAR MATERIAL.

"ABOUT 99% OF MATTER WAS THOUGHT TO BE FORMED BY AN EQUAL AMOUNT OF *HYDROGEN* AND *HELIUM*, WITH THE REMAINING 1% COMPRISED BY THE HEAVIER ELEMENTS.

"IT WAS NATURAL TO ASSUME THAT THE OBSERVED ABUNDANCES OF THE CHEMICAL ELEMENTS DON'T RESULT FROM *NUCLEOSYNTHESIS* IN STARS, WHICH WOULD LEAD TO A *VARIETY* OF CHEMICALS, BUT GO BACK TO THE UNIVERSAL STATE WHEN MATTER HAD YET TO COALESCE INTO STARS AND WAS DISTRIBUTED HOMOGENOUSLY IN SPACE.

"ACCORDING TO *PROFESSOR FRIEDMANN'S* ORIGINAL THEORY OF THE EXPANDING UNIVERSE, BASED ON EINSTEIN'S EQUATIONS OF GENERAL RELATIVITY, THE UNIVERSE STARTED WITH A *"SINGULAR STATE"* AT WHICH THE DENSITY AND TEMPERATURE OF MATTER WERE PRACTICALLY *INFINITE*.

"ATOMS OR ATOMIC NUCLEI DID NOT YET EXIST AND EQUAL NUMBERS OF PROTONS, ELECTRONS, AND NEUTRONS MERGED INTO THE OCEAN OF *HIGH ENERGY RADIATION*.

"I CALL THIS COSMIC SOUP *'YLEM'*. THE WORD, I THINK, COMES FROM A MIDDLE ENGLISH WORD FOR *"MATTER"* OR SOMETHING. I DON'T REMEMBER EXACTLY, BUT I THINK ITS ROOTS WERE IN ANCIENT *HEBREW* OR *GREEK*.

"ANYHOW, AS THE UNIVERSE EXPANDED AND COOLED, PROTONS AND NEUTRONS MUST HAVE BEGUN TO STICK TOGETHER, FORMING *DEUTERONS*--STABLE PROTON-NEUTRON PAIRS--AND THUS THE HEAVY *HYDROGEN NUCLEI*.

"WHEN A NEUTRON *COLLIDES* AND *MERGES* WITH A NUCLEUS, IT DOES SO MORE EASILY THAN A POSITIVELY CHARGED PROTON CAN, BECAUSE IT HAS NO ELECTROSTATIC CHARGE TO BE REPELLED. THIS BUILDS A HEAVIER NUCLEUS, AND IS CALLED *NEUTRON CAPTURE*.

"AS FURTHER CLUSTERING MUST HAVE LED TO HEAVIER AND HEAVIER NUCLEI, THE FINAL RESULT IS THE PRESENTLY OBSERVED ABUNDANCES OF VARIOUS CHEMICAL ELEMENTS."

"IF YOU KNOW THE *PROBABILITIES* OF NEUTRON-CAPTURE BY DIFFERENT NUCLEI YOU CAN CALCULATE THE EXPECTED AMOUNT OF VARIOUS ATOMIC SPECIES AND COMPARE THEM WITH THE OBSERVED DATA.

"SO, IN 1946, AT THE *APPLIED PHYSICS LABORATORY* AT *JOHNS HOPKINS UNIVERSITY*, I WAS JOINED BY *RALPH ALPHER*, A GRAD STUDENT OF MINE, AND *ROBERT HERMAN*, AN EMPLOYEE OF THE LAB, IN A QUEST TO LEARN THE ORIGIN OF THE CHEMICAL ELEMENTS OF THE UNIVERSE.

"CONSIDERING THE BEHAVIOR OF THE EXPANDING UNIVERSE RIGHT AFTER THE 'BIG BANG', WE CONCLUDED THAT AT THAT TIME *THERMAL RADIATION* PLAYED A MORE VITAL ROLE THAN THAT OF MATERIAL PARTICLES. THE MASS DENSITY OF RADIATION MUST HAVE BEEN MUCH LARGER THAN THE COMBINED MASS OF ALL MATERIAL PARTICLES.

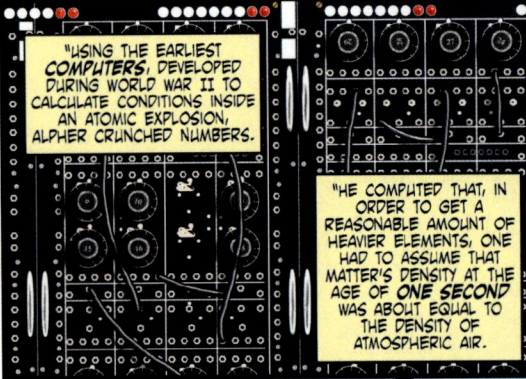

"USING THE EARLIEST *COMPUTERS*, DEVELOPED DURING WORLD WAR II TO CALCULATE CONDITIONS INSIDE AN ATOMIC EXPLOSION, ALPHER CRUNCHED NUMBERS.

"HE COMPUTED THAT, IN ORDER TO GET A REASONABLE AMOUNT OF HEAVIER ELEMENTS, ONE HAD TO ASSUME THAT MATTER'S DENSITY AT THE AGE OF *ONE SECOND* WAS ABOUT EQUAL TO THE DENSITY OF ATMOSPHERIC AIR.

"EXTRAPOLATING FROM THE EARLY DAYS OF THE UNIVERSE TO THE PRESENT TIME, WE FOUND THAT DURING THE PASSING EONS, THE UNIVERSE MUST HAVE COOLED TO ABOUT *FIVE DEGREES* ABOVE ABSOLUTE ZERO, WHICH ROUGHLY AGREED WITH THE WELL-KNOWN TEMPERATURE OF SPACE TODAY, *2.725 KELVIN.**

*THE KELVIN SCALE IS IDENTICAL TO THE CELSIUS SCALE, BUT STARTS AT *ABSOLUTE ZERO* (-273.15°C OR -459.67°F), THE TEMPERATURE AT WHICH THE OSCILLATION OF ATOMS AND MOLECULES ARE AT THEIR VERY SLOWEST.

"IN 1948 WE PUT OUR FINDINGS IN A PAPER FAMOUSLY NICKNAMED THE '*ALPHER-BETHE-GAMOW*' PAPER.

"EVEN THOUGH HANS BETHE DID NOT A LICK OF WORK ON THE PROJECT, I THOUGHT IT WOULD BE FUNNY TO PUT HIS NAME ON THE PAPER SO WE'D GET A GOOD PUN ON '*ALPHA-BETA-GAMMA*'.

The Origin of Chemical Elements

R. A. ALPHER*
Applied Physics Laboratory, The Johns Hopkins University, Silver Spring, Maryland

AND

H. BETHE
Cornell University, Ithaca, New York

AND

G. GAMOW
The George Washington University, Washington, D. C.
February 18, 1948

"UNFORTUNATELY--AND I FEEL BAD ABOUT THIS--USING BETHE'S PRESTIGIOUS NAME AS A *JOKE* UNINTENTIONALLY *OVERSHADOWED* ALPHER, WHO DESERVED MUCH MORE CREDIT THAN THE GOT FOR THE WORK.

"WHILE OUR CALCULATIONS *DID* BASICALLY CORRESPOND TO SCIENTIFIC OBSERVATIONS OF STARS--*HYDROGEN* AND *HELIUM* COMPOSING ABOUT 99% OF THE MATTER IN THE UNIVERSE--OUR STUDY FELL SHORT OF EXPLAINING THE *REST* OF THE CHEMICAL ELEMENTS. BUT, HEY--WE GOT *99%* RIGHT, RIGHT?

"BUT THIS WAS SOON REMEDIED WHEN ASTROPHYSICISTS LEARNED THAT THE OTHER ELEMENTS WERE FORMED NOT AT THE ORIGIN OF THE UNIVERSE, BUT IN STELLAR INTERIORS.

"SO, THOUGH OUR THEORY WAS NOT "*WRONG*", IT WAS INCOMPLETE.

"EARLY STARS WERE **MASSIVE**, CONSUMING HYDROGEN, HELIUM AND LITHIUM TO PRODUCE HEAVIER ELEMENTS.

"THEY SOON DIED, HOWEVER, THEIR **SUPERNOVA DEATH THROES** SEEDING SPACE WITH HEAVIER ELEMENTS SUCH AS CARBON AND OXYGEN, WHICH EVENTUALLY CONDENSED AND FORMED NEW STARS AND PLANETS.

"BUT THIS THEORY OF THE UNIVERSE'S ORIGIN HAD NOT YET WON THE RACE! IT RAN NECK-AN-NECK WITH A VERSION OF THE STEADY-STATE THEORY THAT HELD THAT THE DENSITY OF THE UNIVERSE STAYED BASICALLY THE SAME DUE TO A CONSTANT CREATION OF MATTER.

"ONE OF THIS STEADY-STATE THEORY'S BIGGEST PROPONENTS WAS **FRED HOYLE**, WHO FAMOUSLY SAID IN 1948...

WE NOW COME TO THE QUESTION OF APPLYING THE OBSERVATIONAL TESTS TO EARLIER THEORIES BASED ON THE HYPOTHESIS THAT ALL MATTER IN THE UNIVERSE WAS CREATED IN A **BIG BANG** AT A PARTICULAR TIME IN THE REMOTE PAST.

IT'S AN **IRRATIONAL** PROCESS THAT **CANNOT** BE DESCRIBED IN SCIENTIFIC TERMS, NOR **CHALLENGED** BY AN APPEAL TO OBSERVATION.

AND THUS WAS THE **"BIG BANG"** THEORY NAMED BY ITS TOUGHEST CRITIC.

AT ONE POINT HE SEEMED TO BE CORRECT-- THE OBSERVED EXPANSION RATE OF THE BIG BANG ONLY GAVE THE UNIVERSE AN AGE OF ONLY A **FEW BILLION YEARS**, FAR YOUNGER THAN THE KNOWN AGE OF THE UNIVERSE, WHICH IS AROUND **13.8 BILLION YEARS**.

BUT, AS IT TURNED OUT, WHEN **HUBBLE** HAD MEASURED THE DISTANCE TO SPIRAL NEBULAE HIS CALCULATIONS WERE **TOO SMALL BY HALF**, AND THE COSMIC DISTANCE SCALE HE HAD THUS ESTABLISHED HAD TO BE ADJUSTED ACCORDINGLY.

"NOW, YOU'LL RECALL THAT ALPHER, HERMAN AND I HAD THEORIZED THAT AN OCEAN OF HIGH ENERGY RADIATION HAD SATURATED THE UNIVERSE IN ITS EARLY DAYS, AND WOULD OVER TIME CONVERT INTO MATTER.

"THE TWO OF THEM HAD PREDICTED IN 1948 THAT A RESIDUAL OF THE RADIATION WOULD SPREAD THROUGHOUT ALL OF SPACE UNTIL THE PRESENT DAY--A **COSMIC MICROWAVE BACKGROUND RADIATION**, OR **CMBR**.

"AT ONE POINT I FOUND MYSELF RUNNING DOWN THE TRACKS OF A SPIRAL OF DNA, WHILE THE HUGE FLOATING HEAD OF *JAMES WATSON* EXPLAINED IT ALL TO ME..."

DNA, RNA, AND PROTEINS ARE THREE *MACROMOLECULES* INCLUDED IN ALL FORMS OF LIFE.

DNA CODES FOR RNA AND PROTEINS, TOO, AND *GENES* ARE WHAT WE CALL THE DNA SEGMENTS OF CODED GENETIC INFO.

WHAT YOU SEE IS A STYLIZED STRUCTURAL DNA *MODEL* INSPIRED BY THE ONE FRANCIS AND I DISCOVERED BACK IN 1953. KEEP THIS IN MIND--YOU WOULD *NOT* ACTUALLY SEE THIS THROUGH AN ELECTRON MICROSCOPE.

THE "RAILS" OF THE SPIRAL LADDER--ITS "BACKBONES"--ARE MADE OF *PHOSPHATE* GROUPS BONDED TO FIVE-ATOM *DEOXYRIBOSE* SUGAR MOLECULES.

ATTACHED TO EACH DEOXYRIBOSE MOLECULE IS ANOTHER MOLECULE CALLED A *NUCLEOBASE*. BASES COME IN FOUR "FLAVORS": *ADENINE*, *CYTOSINE*, *GUANINE* AND *THYMINE*.

EACH BASE IS VERY PICKY ABOUT WHICH OTHER BASE IT'LL "HOOK UP"-- ADENINE PAIRS ONLY WITH THYMINE AND CYTOSINE ONLY WITH GUANINE. A *HYDROGEN BOND* HOLDS THEM TOGETHER LIKE A NEWBORN INFANT.

THE PHOSPHATE-SUGAR-BASE TRIO IS CALLED A *NUCLEOTIDE*. THE NUCLEOTIDE SEQUENCE ALONG THE BACKBONES IS A CODED GENETIC "TEXT" SPECIFYING THE AMINO ACID SEQUENCE WITHIN PROTEINS.

GOOD OL' *GEORGE GAMOW* WAS THE FIRST TO POSTULATE THE CODON'S TRIPLICATE NATURE!

EACH SERIES OF *THREE ADJACENT BASES* IN A POLYNUCLEOTIDE LINK OF DNA OR RNA IS CALLED A *CODON* BECAUSE IT CODES FOR ONE SPECIFIC AMINO ACID.

HAHAHA! SUCH *STRANGE* DREAMS YOU HAVE, TOMPKINS!

BUT YES, JIM WATSON WAS RIGHT. I *WAS* THE FIRST TO POSTULATE THE CODON'S TRIPLICATE NATURE. IT'S QUITE APPROPRIATE THAT I WAS PLAYING CARDS IN YOUR DREAM, ACTUALLY...

A WATSON-CRICK PAPER IN A 1953 ISSUE OF *NATURE* DISCUSSED HOW HEREDITARY INFORMATION IS STORED IN DNA MOLECULES IN THE FOUR NUCLEOBASES.

AFTER READING IT I WONDERED HOW THIS INFORMATION TRANSLATED INTO THE SEQUENCE OF 20 AMINO ACIDS FORMING PROTEIN MOLECULES.

THEN IT OCCURRED TO ME THAT YOU *CAN* "GET 20 OUT OF 4" BY COUNTING THE NUMBER OF ALL POSSIBLE *TRIPLETS* FORMED OUT OF FOUR DIFFERENT ENTITIES.

HERE, TAKE THIS DECK OF *CARDS*.

EACH OF THE FOUR SUITS REPRESENTS ONE NUCLEOBASE.

NOW TELL ME--HOW MANY TRIPLETS ALL OF THE SAME SUIT CAN YOU GET?

UMM... *FOUR*. THREE HEARTS, THREE DIAMONDS, THREE SPADES, AND THREE CLUBS.

OKAY, NOW HOW MANY TRIPLETS WITH *TWO* CARDS OF THE SAME SUIT, AND ONE DIFFERENT?

WELL, WE HAVE *FOUR* CHOICES FOR THE PAIR, AND... *THREE* CHOICES FOR THE THIRD CARD.

SO, *FOUR* TIMES *THREE* EQUALS TWELVE. TWELVE POSSIBILITIES.

"Also—this just occurred to me—I have YOU to thank for introducing me to my new wife, PERKY."

"Her name is 'PERKY'?"

"Well, that's what I call her! Listen!"

"In the summer of '53, I had to sail to ENGLAND for a meeting."

"The manager of the American branch of CAMBRIDGE UNIVERSITY PRESS was to see me off at the dock and give me a copy of my newest book, MR. TOMPKINS LEARNS THE FACTS OF LIFE, as well as a bottle of scotch whisky."

"Well, he couldn't make it, so in his stead stood the company's lovely publicity manager—MISS BARBARA PERKINS!"

WELL, I WISH YOU MANY HAPPY YEARS TOGETHER. AND NOW IT'S THAT TIME WHERE I PULL MY DISAPPEARING ACT.

SO LONG, "DAD". I LOOK FORWARD TO OUR NEXT MEETING.

LIKEWISE, "SON". SAY, WOULD YOU MIND IF I WATCH AS YOU WORK THAT TIME MACHINE? I'D BE DISAPPOINTED TO HAVE HAD THE OPPORTUNITY BUT NOT TAKEN IT.

I PROMISE I WON'T LOOK TOO CLOSELY AT THE DETAILS AND THUS CHANGE HISTORY.

SEE YOU IN THE FUNNY PAPERS!

AMAZING!

"MY PREVIOUS WIFE, LYUBOV ('RHO' AS I CALLED HER), AND I HAD RECENTLY DIVORCED, SO OVER THE NEXT FIVE YEARS BARBARA AND I GOT TO KNOW EACH OTHER WHENEVER I WOULD VISIT NEW YORK. IN 1958 WE MARRIED!"

AFTERWORD
BY IGOR GAMOW

As you now can see from having read Father's life story, he was quite a character, and a preeminent storyteller. I profess to have inherited the storytelling gene, so as this book draws to a close I would like to share a humorous personal reminiscence of my time spent in one of Father's classes.

I was a freshman at the University of Colorado in 1958, and enrolled in my father's undergraduate physics course for non-majors. The class was called "Matter, Earth, and Sky", and Father used his then-new book of the same name to teach it. This was the first time I had been in a classroom with Father. There were about a hundred students in that course, and I bet they all well-remember being there.

In this class Father decided, for the first time, to demonstrate his own experiments. This was a bit surprising, for he had been an uninspired experimentalist while a graduate student at the University of Leningrad. Perhaps, then, he simply thought he would give experimentation another try. The plan was to demonstrate an experiment attributed to Archimedes of Syracuse, an ancient Greek scientist.

As the story goes, according to Book IX of Vitruvius' *de Architectura*, Archimedes was asked by King Hiero II of Syracuse to determine whether the votive gold crown made for him to be placed in a temple to the gods was truly solid gold or if any silver had been used in its construction by the potentially "shifty" blacksmith. Since the crown was consecrated to the gods, Archimedes was not allowed to scratch, deface, or melt the crown.

The great scientist wasn't sure what to do. One day, however, as he stepped into a bath (perhaps at the court of Hiero II), he noticed that the more his body sank into it, the more water flowed over the rim of the bathtub.

"Eureka!" he shouted in Greek (ευρηκα, mearing "I have found it!"). Leaping from the tub, he ran home, naked, repeating his exclamation all the way.

Next, as Vitruvius relates:

> (Archimedes) made two masses of the same weight as the crown, one of gold and the other of silver. After making them he filled a large vessel with water to the brim, and

dropped the mass of silver into it. As much water ran out as was equal in bulk to that of the silver sunk in the vessel. Then, taking out the mass, he poured back the lost quantity of water... until it was level with the brim as it had been before. Thus he found the weight of silver corresponding to a definite quantity of water.

After this experiment, he likewise dropped the mass of gold into the full vessel and, on taking it out and measuring as before, found that not so much water was lost, but a smaller quantity: namely, as much less as a mass of gold lacks in bulk compared to a mass of silver of the same weight. Finally, filling the vessel again and dropping the crown itself into the same quantity of water, he found that more water ran over for the crown than for the mass of gold of the same weight. Hence, reasoning from the fact that more water was lost in the case of the crown than in that of the mass, he detected the mixing of silver with the gold, and made the theft of the contractor perfectly clear.

Such was how Archimedes supposedly measured the gold crown. But measuring a gold crown at the University of Colorado was a different matter. Since the high price of gold obviously precluded CU from financing the creation of a real gold crown for the demonstration, its machine shop instead constructed one of bronze. On the day of the lecture, Father came into class with the "gold" crown and explained how he was going to measure the volume of water displaced when the crown was immersed in a beaker of water.

So far, so good.

With a string he dangled the crown from a ring stand and put a 5000 cc glass beaker full of water beneath them. His idea was pretty simple--he would lower the crown into the beaker of water and measure how much water was displaced.

Explaining the principle to the class he said, "Now I will lower the crown into the water."

He turned the thumbscrew on the ring stand. The crown crashed down into the beaker, smashing it. Glass shards and water went everywhere, even splashing the students in the front row beside me.

Father was soaked from the waist down. The students howled with laughter.

"Well," Father said meekly, "this is an experiment in dynamics, not in density."

Ever-persistent, he wondered what he could do to save the experiment. Though all the water was gone, a sink was built into the lecture table. There was the answer.

"The day is saved!" he triumphantly announced. "We will submerge the crown into the water in the sink."

Reaching down, he turned on the faucet, releasing not a stream of water but a cloud of steam which rose from the basin and completely fogged his glasses. A girl next to me laughed so hard that she began to heave, and the other students stomped up a tremendous cacophony. As Father finally turned off the nozzle I tried to make myself as small as possible.

Emerging from the steam, Father said, "Oh, wrong faucet."

Finding the correct knob, he said, "Ah, water," and turned it.

But the water faucet had a long rubber hose attached, and it writhed like a sea serpent out of the sink and sprayed water everywhere.

"Turn off the water!" yelled the students. "Turn off the water!"

By this time the physics department secretaries, as well as other people milling about in the hallway outside the classroom, poured in. Father's assistant, Dalton, ran to the rescue. Taking Father by the elbows, he moved him aside and cleaned up the mess with a mop and broom.

Clutching a ream of wet notes, Father announced. "The experiment is over."

Wolfgang Pauli
(1900-1958)

Father was convinced that the better you are as a theoretical physicist the worse you are as an applied physicist. "It is well known," he wrote in *Thirty Years That Shook Physics*, "that theoretical physicists cannot handle experimental equipment; it breaks whenever they touch it." This phenomenon was jokingly called "The Pauli Effect", and Father loved to tell the story of how it received its name. One day in Göttingen, Germany, some atomic machinery in the laboratory of physicist James Franck collapsed without apparent reason. Franck, perplexed, wrote about this occurrence to Wolfgang Pauli, probably the world's greatest theoretical physicist at that time. Pauli wrote back, explaining that he had taken a train to visit Niels Bohr in Denmark, and at the moment the machinery had broken the train had stopped at the Göttingen railway station.

Thus arose the humorous notion that if Pauli were nearby, scientific apparatus would cease functioning. Since Pauli died in the last month of 1958, the very year my father's experiment took place, one might wonder if, before his death, Pauli might not have been visiting Boulder on the day Father's gold crown experiment went awry. However, I would certainly have seen him if he had visited the University of Colorado and Boulder, so such was not the case. So perhaps Father was right, and the Pauli Effect is a phenomenon common to all theoretical physicists, as he stated. To be sure, the Pauli Effect was indeed in full force on the day of Father's experiment, yet despite it all, Father still persisted to perform his own demonstrations throughout the course...

And they occasionally (kind of) worked.

A lesson to learn from all this is that one should be tenacious in one's endeavors. Never give up, especially in pursuit of knowledge. One might think, for example, that Mr. Tompkins would be afraid to go to sleep at night, as most of his dreams end up in scientifically related disasters. But in spite of this, Tompkins' desire for learning outweighs any nightmares or temporary setbacks which arise from his love of science. As he has now endured eighty years, Mr. Tompkins is an example to us all.

Igor Gamow
Boulder, Colorado
December, 2016

ALSO AVAILABLE FROM

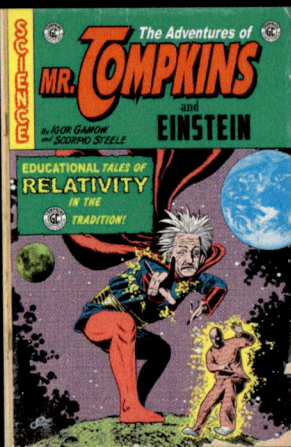

EINSTEIN
ISBN 978-1477575772
ALSO AVAILABLE FOR KINDLE

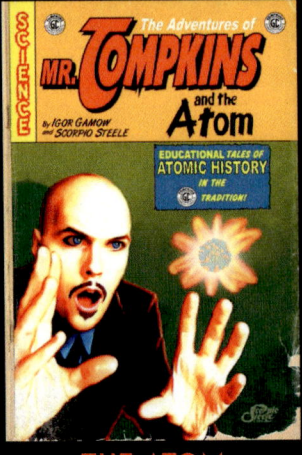

THE ATOM
ISBN 978-1477575888
ALSO AVAILABLE FOR KINDLE

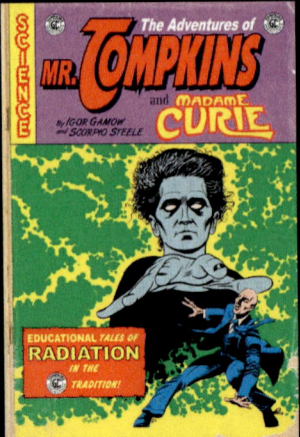

MADAME CURIE
ISBN 978-1477575710
ALSO AVAILABLE FOR KINDLE

DARWIN
ISBN 978-1477453933
ALSO AVAILABLE FOR KINDLE

MENDEL
ISBN 978-1477575895

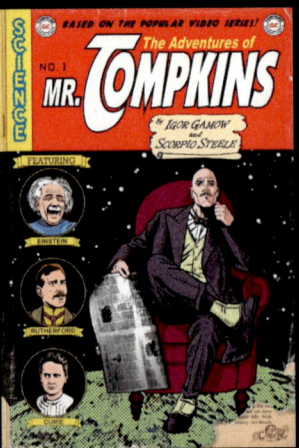

VOLUME ONE
COLLECTS EINSTEIN,
THE ATOM, AND CURIE
ISBN 978-1439252116

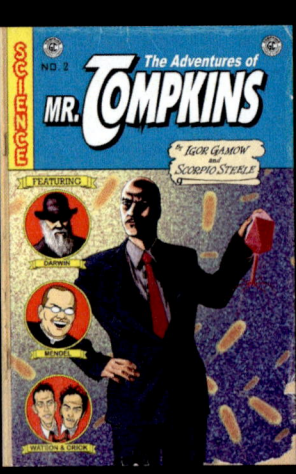

VOLUME TWO
COLLECTS DARWIN,
MENDEL, AND DNA
ISBN 978-1461195597

BIG BANG PRODUCTIONS

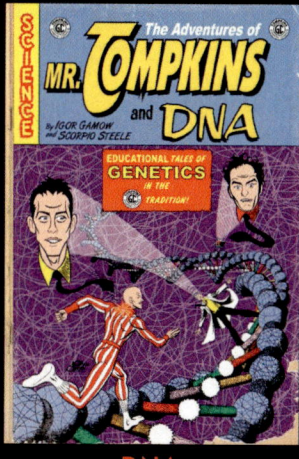

DNA
ISBN 978-1477575925

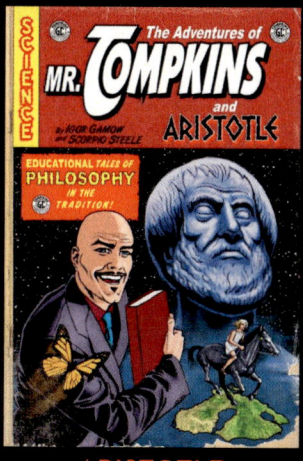

ARISTOTLE
ISBN 978-1482375183

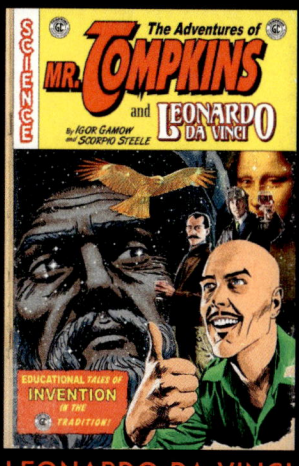

LEONARDO DA VINCI
ISBN 978-1492365273

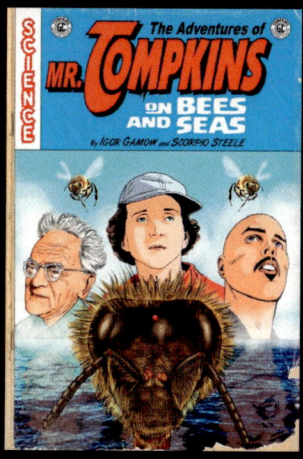

ON BEES AND SEAS
ISBN 978-1495257254

GEORGE GAMOW
ISBN 978-1502854612

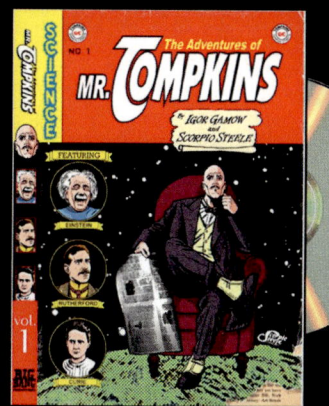

DVD
FEATURES LECTURES,
PLUS VIDEOS OF EINSTEIN,
THE ATOM, AND CURIE
ASIN: B0032AM8LY

GEORGE GAMOW was a cosmologist and theoretical physicist who helped formulate and popularize the Big Bang theory of the origin of the universe. He developed a theory of the mechanism of alpha decay via quantum tunneling, studied radioactive decay of the atomic nucleus, stellar nucleosynthesis, Big Bang nucleosynthesis, and molecular genetics. Gamow was also a prolific author, writing more than twenty books, including the popular *Mr. Tompkins series*. More than fifty years after original publication, many of his books are still in print.

IGOR GAMOW dropped out of high school to become a dancer with the National Ballet Company in Washington, D.C. At age twenty-two Igor returned to academia at the University of Colorado, and his research eventually led to numerous inventions in the field of bionics, including the Gamow Bag, a portable hyperbaric chamber for high-altitude climbing. With artist Scorpio Steele, Igor authors the educational comic book series *The Adventures of Mr. Tompkins*, based on the award-winning books created by his father, George Gamow.

SCORPIO STEELE stayed in high school and graduated. He attended the University of Colorado with hopes that he might learn some useful art techniques in his art classes. He did not, and still considers himself 100% self-taught. Occasionally he goes dancing, and he enjoys watching Steve Austin, The Six Million Dollar Man, use his bionics. The only ballet he has ever seen live is *The Nutcracker*, and he has still never been to Washington, D.C. At age twenty-two his first comic book was published, but he won't tell you the name of it unless you get to know him well. Since 2008 he and Igor Gamow have been creating *The Adventures of Mr. Tompkins*, based on the award-winning books created by Igor's father, George Gamow.

PAUL D. BEALE is a professor of physics at the University of Colorado Boulder. A theoretical physicist specializing in the statistical mechanics of condensed matter systems, his research interests include exact solutions of statistical mechanical models, the thermodynamics, phase transitions, and critical properties of materials, and pseudorandom number generators. With R.K. Pathria he coauthored the third edition of *Statistical Mechanics*, a graduate physics textbook. He wrote the forward to *The Complete Adventures of Mr. Tompkins*.

Made in the USA
Middletown, DE
04 January 2026

26623473R00168